Crimeproofing Your Business

Crimeproofing Your Business

301 Low-Cost, No-Cost Ways to Protect Your Office, Store, or Business

Russell L. Bintliff

McGraw-Hill, Inc.

New York   San Francisco   Washington, D.C.   Auckland   Bogotá
Caracas   Lisbon   London   Madrid   Mexico City   Milan
Montreal   New Delhi   San Juan   Singapore
Sydney   Tokyo   Toronto

Library of Congress Cataloging-in-Publication Data

Bintliff, Russell L.
    Crimeproofing your business : 301 low-cost, no-cost ways to
protect your office, store, or business / Russell L. Bintliff.
            p.   cm.
    Includes index.
    ISBN 0-07-005308-1 (hc : acid-free paper)—ISBN
0-07-005309-X (pbk. : acid-free paper)
    1. Industry—Security measures—United States.   I. Title.
HV8291.U6B56   1994
658.4′73—dc20   93-38009
                                                                CIP

1 2 3 4 5 6 7 8 9 0   DOC/DOC   9 0 9 8 7 6 5 4

ISBN 0-07-005309-X (pbk)
ISBN 0-07-005308-1 (hc)

*The sponsoring editor for this book was Caroline Carney, the editing supervisor
was Ruth W. Mannino, and the production supervisor was Suzanne W. Babeuf.
It was set in Palatino by McGraw-Hill's Professional Book Group composition
unit.*

Printed and bound by R. R. Donnelley & Sons Company.

To My Family

Contents

Preface   xi

1. Assessing the Vulnerability of Your Business   **1**

Evaluating Your Business Exterior   2
Analyzing Your Business Interior   10
Testing Your Management Controls   13
Quick Reference Guide to Chapter 1: Assessing Your Threats   16

2. Choosing the Best Locks for Your Business   **18**

Six Common Types of Business-Related Locks   18
The Pros and Cons of Lock Security   24
The Art and Science of Lock Picking   31
Controlling Locks and Keys   33
Designing a Key Control System   35
Quick Reference Guide to Chapter 2: Installing a Lock and Key
  System   42

3. Increasing Structural Security   **44**

Hardening the Entry Points of Your Business   44
Hardening Exterior Walls and Roofs on Your Business
  Premises   50
Protective Lighting: The Overlooked Technique   56
Types of Protective Lighting Systems   57
The "Best" Lighting for Protecting Your Business Premises   59

Protective Lighting Maintenance   61
Quick Reference Guide to Chapter 3: Protective Exterior
  Lighting   63

**4. Choosing the Right High-Tech Security Devices for
   Your Business**   **65**

What Is an Intrusion Detection System?   65
Benefits and Advantages of an IDS   66
Factors Affecting Application of an IDS   67
Basic Parts of an Intrusion Detection System   70
Problems with Intrusion Detection Systems   71
Solutions to IDS Problems   73
Overextension of IDS Equipment   74
Sound-Activated Detection Devices   82
Other Detection Devices   88
Quick Reference Guide to Chapter 4: Six Key IDS Factors   94

5. Eliminating the Threat of Armed Robbery   **95**

Armed Robbery: A Definition   96
The Anatomy of Armed Robbery   98
The Business of Armed Robbery   103
Assessing Entry and Escape Routes   105
Quick Reference Guide to Chapter 5: Eliminating Business
  Robberies   110

6. Putting an End to Employee Pilferage   **112**

Understanding the Magnitude of Employee Pilferage   113
Why Do Employees Steal?   121
Inventory Accountability and Control   126
Employee Pilferage Against Business Customers   132
Quick Reference Guide to Chapter 6: Preventing Employee
  Pilferage   134

7. Eliminating Bad Checks and Credit Card Fraud   **136**

The Bad Check Artist: A Profile   136
The Scope of Credit Card Fraud   150
Quick Reference Guide to Chapter 7: Paper and Plastic
  Fraud   152

**8. Spotting Counterfeit Cash and Bogus Discount
   Coupons**   **154**

The Mechanics of Counterfeit Currency   154
Types of Paper Currency   156
How Criminals Counterfeit Currency   159

How Counterfeit Money Circulates   162
The Other Currency: Discount Coupons   163
Coupon Kings and Queens   168
Quick Reference Guide to Chapter 8: Detecting Counterfeit
  Currency   172

9. Ending the Shoplifting Problem   174

A Time-Honored Profession   174
The Return Merchandise Scam   186
Catching Shoplifters in the Act   188
Quick Reference Guide to Chapter 9: The Shoplifter's Bag
  of Tricks   191

10. Avoid Becoming a Victim of Business-to-Business Fraud   194

The Pervasive Gray Market   194
Duplicate-Billing Schemes   197
The Office Supply Scam   199
Unfair Methods of Competition   201
Violations of Antitrust Law   205
Antitrust Remedies and Enforcement   212
Quick Reference Guide to Chapter 10: Key Concepts in
  Business-to-Business Fraud   214

11. Safeguarding Your Business Information   216

How Secure Is Your Business Information?   217
The Perils of Public Places   221
Internal Security Measures   223
The Vulnerability of Mobile Offices   227
Information Embezzlement and Computer Crime   229
Quick Reference Guide to Chapter 11: Safeguarding
  Computerized Business Information   235

12. Preventing the Diversion of Business Cargo   237

Protective Management of Intransit Cargo   238
The Ripple Effect of Cargo Losses   244
Transportation Mode Evaluation Guide   246
Passive Techniques to Prevent Cargo Loss   250
Quick Reference Guide to Chapter 12: Protecting
  Your Cargo   252

13. The Fine Art of Hiring Employees   253

Security Begins with the Hiring Process   253
Critical Steps in Preemployment Screening   256

Processing the Employment Application   259
Tests to Screen Applicants   262
The Pros and Cons of Applicant Interviewing   264
Quick Reference Guide to Chapter 13: Screening Job
  Applicants   266

**14. Training Employees: The Secret Weapon of Asset
Protection**   **267**

The Critical Importance of Employee Training   268
A Performance-Oriented Approach to Training   269
How to Create a Business Training Session   273
How to Evaluate Your Employee Training Program   278
Creating a Positive Training Environment   280
Quick Reference Guide to Chapter 14: Training Employees
  Effectively   281

15. Hiring Security Officers or Consultants   **283**

Security Officers and Consultants: Some Key Definitions   283
Contracting for Private Security Officers   288
Choosing and Hiring an Asset Protection Consultant   291
Quick Reference Guide to Chapter 15: Hiring Security Officers or
  Consultants   297

Appendixes   **299**

A The Businessperson's Observation, Description, and
     Identification Guide   301
B Physical Security Checklist for Retail Stores   323
C Physical Security Checklist for Retail Grocery Stores   327
D Physical Security Checklist for Warehouse Activities   331
E Physical Security Checklist for Finance and Accounting
     Centers   333
F Checklist for Preventing White-Collar and Economic Crime   336
G Checklist for Planning Inventory Controls   346
H Criteria for Data-Processing Safeguards   355
I Physical Security Plan for Businesses   359

Index   365

Preface

You know the importance of planning for your business, but have you considered how to protect your business assets? Crime, targeting businesses of all sizes, has become one of the fastest growing and most lucrative industries, reaping illicit profits in excess of $200 billion a year. Some analysts believe that the losses are double that amount because most businesses do not report the crimes. Often businesses do not recognize their losses as stemming from criminal enterprises. Other times, they believe it goes with the territory, that there's nothing they can do about it.

The truth is that you do not have to sustain business losses from these crimes. All crime, including those targeting your type or place of business, has one thing in common—opportunity. Without opportunity a crime cannot happen. To protect your business and its assets, you need to eliminate opportunity.

Throughout this book I will show you how to make security a routine part of your daily management and operations. More importantly, I'll show you how eliminating criminal erosion dramatically increases your net profits.

The best way to start is to ignore any security measures currently in place. After designing the new systems protection, you can assess the existing measures to determine if any of them complement the new program. We proceed in this way because existing measures often either do not work or need significant modification to meet security standards. Thus, each of this book's chapters shows you how to start from ground zero. By the end of the book, you will have in place an asset protection management program where none existed, or a recon-figuration of your existing program. In either case, this book will enable you to start running a more secure business right away.

As you scan through the book, you will notice 301 different security techniques. Each of these methods is an affordable, or even no-cost, way to reduce your vulnerability to crime. You might be tempted to pick and choose a couple of these techniques and ignore the rest. You are welcome to do so. However, I would caution you against taking half measures. After 30 years of training and experience, I can assure you that the only approach that supplies you with real results is to develop an overall, operations friendly system. This "total system approach" consists of mutually supporting layers of preventive ele-ments coordinated to prevent gaps or overlap and to ensure tight secu-rity. A complete reading of this book will enable you to set up this total system. It will help you to create a crimeproof business.

Russell L. Bintliff

1
Assessing the Vulnerability of Your Business

Don't forget that you're in business to make money. In any business, that begins with absolute control of all your assets—money, product, merchandise, equipment, and other items that create your investment. Your business asset protection program must begin with a realistic examination of vulnerability. For example, a warehouse differs importantly from a retail outlet or restaurant, while an electronics supplier or manufacturer differs importantly from a company that distributes cardboard boxes. Each has vulnerability at different levels and often to different criminal enterprise schemes. However, a common denominator runs through all businesses regardless of type, size, or location: Whatever assets the business possesses, someone will always try to steal them, either overtly by robbery or burglary or, more often, covertly by shoplifting, embezzlement, or pilferage.

To thwart losses from these overt and covert criminal activities, you first need to assess your business vulnerability and establish a protective foundation. Without it, you cannot implement effective, meaningful, or lasting countermeasures, any more than you can build a house without a solid foundation.

Individual circumstances, such as location, type of building, crime statistics (in the city, town, suburban, or rural location), value of property, and value of contents affect your business vulnerability. The assessment techniques in this book have been divided into categories that you can adapt to any situation. Included are techniques that apply at the highest threat levels and that can be adjusted to your asset protection needs.

Throughout the book, I will supply you with viable solutions to all the problems cited in this chapter. However, first you need to assess the problems carefully and develop an awareness of your vulnerabilities.

Evaluating Your Business Exterior

Burglary remains a favorite technique of criminal enterprises targeting businesses. It harms your business by depriving you of operational assets, such as merchandise, equipment, products, and information. Burglary also leads to a loss of income, because it diminishes your ability to conduct business, and it often entails structural damage to your building that requires costly repairs. Thus, you need to begin your planning with a structural assessment, and it's best to work from the outside in. Chapter 2 addresses ways of correcting or safeguarding any structural deficiencies you find.

Security Technique 1: Assess the Exterior Walls of Your Business Location

Whether your business is housed in a freestanding structure or is part of a large building or mall, you need to determine the ease of access through the walls. Many businesses have alarm systems connected to doors and windows. Knowing this, professional burglars can readily enter, steal assets, and leave undetected by cutting through a thin wall. Years ago, smaller banks in rural areas and branch banks in suburban areas experienced a huge theft problem. Burglars learned that despite impressive vault doors, many bank vaults had at least one outside wall, normally brick with an inside plaster covering. With relative ease, a burglar could chip out the brick and knock a hole through the plaster, enter the vault, and steal valuables without setting off the alarm system, which relied on entry through the building and vault door. New regulations have put a stop to the burglaries through a variety of countermeasures to strengthen the vaults.

You need to assess your business vulnerability from walls surrounding a freestanding building or from those separating one office, hallway, or storage closet from another office complex. Skilled burglars can easily hide in a hallway storage closet or in the basement of a large office building until after hours, thereby avoiding entrance security problems, and then cut through a plasterboard wall to enter a business. The burglars then steal information, office equipment, computer records and programs, typewriters, and other items, and store them in another location within the building. After completing their theft, the burglars go into hiding until the following business day and slip out with the crowds of people entering and leaving. Later, after the excitement of the theft fades, the burglars return under another guise during business hours and collect their plunder. See Fig. 1-1 for examples of external wall vulnerability.

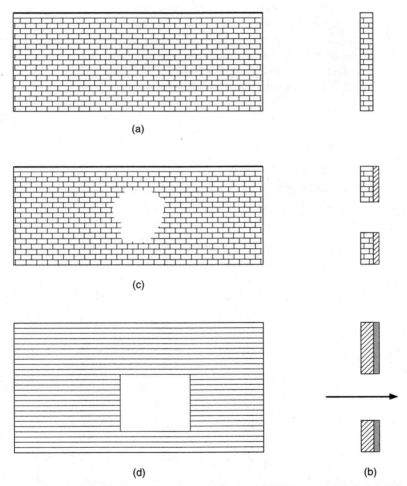

(a)

(c)

(d)   (b)

Figure 1-1. Exterior wall vulnerability. (*a*) Wall appears sturdy. (*b*) Side view of wall, showing its vulnerability. (*c*) With a hammer and chisel, a burglar can quickly create a hole in a seemingly sturdy, average brick and plasterboard wall. (*d*) With a drill and saw a burglar can easily cut through the average wooden exterior wall and plaster/plasterboard backing. Note: The same vulnerability applies within high-rise buildings where burglars can enter from an adjacent room or office, a hallway, or a hall storage closet.

Security Technique 2: Assess the Windows at Your Business Premises

Windows supply a major vulnerability to burglary and robbery. Burglars will view windows, especially those not protected with alarm systems, as an easy way of entering either a freestanding building or an office complex. Breaking door windows is also an easy way of getting to inside locks, especially at night through doors that lead into the building from alleys or other darkened areas. Although seemingly secure, exterior doors usually have a window for viewing persons requesting entry. A skilled burglar can reach through with a hand or tool to open the interior dead bolt or unlock the bottom of a doorknob lock (Fig. 1-2*a*).

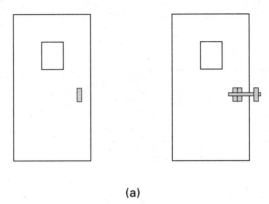

(a)

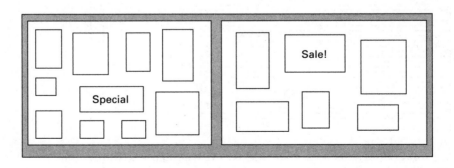

(b)

Figure 1-2. Business window vulnerabilities. (*a*) Outside (*left*) and inside (*right*) view of exterior door with small window. (*b*) A store window covered with advertising posters.

Retail businesses often invite armed robberies by posting large advertising signs on front and side windows encouraging the public to shop there (Fig. 1-2b). Besides informing passersby about sales or store prices, these large posters obstruct the view into the business. A study of rural, suburban, and urban businesses across the country shows that armed robberies occur more frequently at businesses with poster-obscured windows and that passersby, including the police, rarely notice what's happening inside. If your business has large windows and you fill them with posters, you're increasing your chances of being robbed.

Obscured windows also present a lucrative target for burglars who rely on the concealment factor to complete their thefts. A burglar can enter a business with poster-obscured windows after dark or during daylight hours when the business is closed (such as on weekends). Even with the interior well-lit after business hours, burglars may pose as cleaners or stocking crews to avoid discovery if a passerby does see them through narrow spaces between the posters.

Basement windows also need to be evaluated to determine if a person can enter or exit through them. Unprotected or inadequately protected basement windows may also invite internal pilferage. An employee, vendor, or maintenance person, for example, might place merchandise or other business assets outside the basement window to pick up later, or hand them to a person waiting outside.

Security Technique 3: Examine Your Business Access Doors

Doors represent a major vulnerability to burglary (Fig. 1-3). Even doors that appear sturdy and secure to the average person may offer easy access to the professional burglar. Installing high-quality locks can help to protect against picking, but you also need to assess the vulnerability of door frames. Can a burglar pry off the frames and release the door lock? Check door hinges as well. Can a person remove the pins that release the door? A glass door or one with a window that can be easily broken creates additional vulnerability.

You need to examine outside doors to determine whether they have a solid construction or a hollow core. A burglar can cut through a hollow-core door in seconds with a sharp knife, often bypassing contact alarms attached to the door that are triggered only when the door opens in a normal way. Decorative door panels also offer vulnerable points of entry, since they are not affected by contact alarm systems. The panels can easily be kicked or pushed out, especially in hollow-core door construction. Transoms over entry doors, a common feature

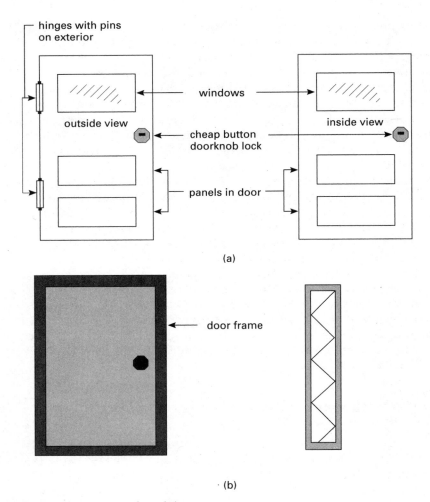

Figure 1-3. Door vulnerabilities.

in older businesses, can become easy access points for burglars. Figure 1-3 shows examples of business door vulnerabilities.

Security Technique 4: Assess the Roof and Ceilings

Business buildings often have a variety of vulnerable entry and exit points on the roof that enable both burglars and robbers to gain surreptitious access (see Fig. 1-4). You need to examine the vulnerability of skylights and decide how easily a person could enter or exit through them. You also need to consider how easily a person can cut through the roof or gain access through heating, cooling, or exhaust ventilators. Sometimes, large air conditioners mounted on the roof can be moved, exposing a large hole underneath. Many high-rise office complexes have an access door and stairs that lead into the building.

In dense urban environments, the proximity of other buildings could allow a person to jump or even walk from one building to another and enter the roof. You need to check whether a roof entrance door has alarm protection and whether it is left unlocked. Also check for vulnerable maintenance roof hatches that allow burglars access to a business through heating or cooling ducts.

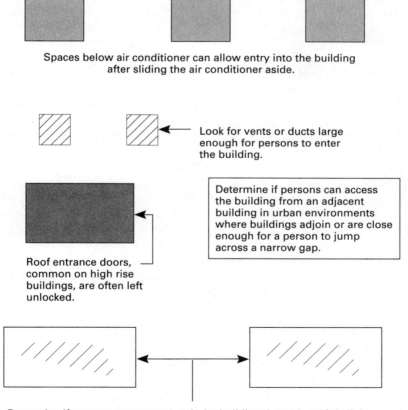

Spaces below air conditioner can allow entry into the building after sliding the air conditioner aside.

Look for vents or ducts large enough for persons to enter the building.

Determine if persons can access the building from an adjacent building in urban environments where buildings adjoin or are close enough for a person to jump across a narrow gap.

Roof entrance doors, common on high rise buildings, are often left unlocked.

Determine if persons can enter or exit the building through roof skylights.

Figure 1-4. Roof vulnerabilities.

Ceilings pose problems in two ways. The first involves "drop ceilings." A few years ago, one of my clients experienced armed robberies in several stores it operated across a southern state. Its managers were being robbed after the stores had closed and all doors were locked. Even though the police came in at closing to check the stockroom, rest rooms, and store floor, the robber conveniently appeared soon after they left. The police and the owners thought that their mysterious robber was a former employee who had a key or some other way of entering the building. But even after they changed the locks and had the building structure inspected for a secret entrance, the robberies continued.

As soon as I looked at the ceilings, I knew exactly what the problem was. In each store, the public rest rooms were required by law to have drop ceilings, with about 4 square ft of ceiling block set on a network of light steel supports. Some 8 ft of space existed between the ceiling and the roof. The robber would simply go into the rest room before the store closed, lift up the ceiling block, pull himself up into the air space, and replace the block. After the store closed, he would reverse the process, drop down into the rest room, replace the ceiling block, walk out into the store, and accost the manager before he or she locked the day's receipts in the safe. After tying up the employee, the robber would exit through a back door into the alley and disappear. By placing security officers in hidden positions within the store and coordinating with the police, we captured the ghost robber the night after my inspection.

Some businesses have drop ceilings that run throughout the store or office complex. A person can easily move along the network of supports to come down anywhere in the premises. Ceiling access could also be gained from a hallway supply closet, from a closet housing a fire hose, or from another office complex. Here again, burglars can easily hide their plunder in the ceiling space in any part of the building until they have an opportunity to remove it.

Later in this book, I will show you how to safeguard drop ceilings as part of your asset protection program.

Security Technique 5: Check Your Outside Lighting

Outside lighting can be an effective and cost-efficient deterrent against burglary. Burglars must operate in secret to stay successful, and they avoid businesses in which passersby, security, or police patrol could detect their presence. Assess your business at night and consider

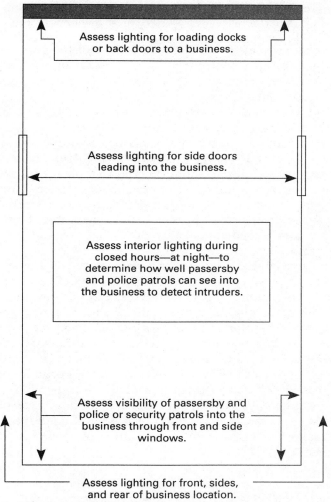

Figure 1-5. Business lighting vulnerabilities.

whether a burglar could disarm outside lights by breaking the elements or cutting power lines that feed electricity to them. Look for lighting that creates shadows or that blinds passersby or other observers. You also need to examine how your lighting illuminates surrounding areas and adjacent buildings. Figure 1-5 shows examples of lighting vulnerabilities to look for in your business.

Analyzing Your Business Interior

Your business interior may also encourage theft. One of the greatest problems all businesses must confront is theft of assets by internal sources (such as employee pilferage) as well as by exterior sources (burglary, robbery, and theft).

How you arrange the interior of your retail business, warehouse, manufacturing plant, office, or other operation sends a signal to people contemplating theft of your assets. For example, if your business involves retail sales, how you arrange the merchandise display areas can either encourage or discourage shoplifting and employee pilferage. Allowing your business storage areas to remain open to employees or vendors might smooth operations, but it also invites opportunities for theft and lost profits. Your interior arrangement might also encourage burglary and armed robbery by providing areas of concealment.

In a warehouse environment, the interior arrangement and flow of incoming and outgoing products might facilitate theft by delivery persons. In a manufacturing environment, employees, vendors, or others having normal access to the plant might find that the interior arrangement helps them to conceal items. In an office environment, equipment and supplies as well as business information files are the important assets to protect. How you arrange your office complex might make the difference between opportunity and denial of opportunity to steal. In urban areas, office complexes are particularly vulnerable to armed robbery, since the physical layout provides numerous areas of concealment. Often robbers can enter, victimize several employees, and then leave with the remainder of the office staff not knowing what happened until later.

Whatever your business, you need to make a thorough assessment of internal vulnerabilities, To help you do that comprehensively, I have divided this important activity into several categories that can easily be adapted to your business situation.

Security Technique 6: Assess Your Retail Store's Interior Vulnerabilities

Retail stores experience staggering losses annually from theft—including shoplifting, burglary, employee pilferage, vendor theft (mostly employees of vendors, such as delivery persons), and armed robbery. The success of these crimes stems directly from opportunity. Despite your best efforts, some crimes will happen. But when your asset protection planning and countermeasures become effective, people who do commit crimes against your business will be quickly identified and arrested or fired, depending on the situation.

Figure 1-6 presents guidelines for assessing your retail store's internal vulnerability. You need to remain attentive to arrangements that conceal or encourage crime. For example, in most retail stores the

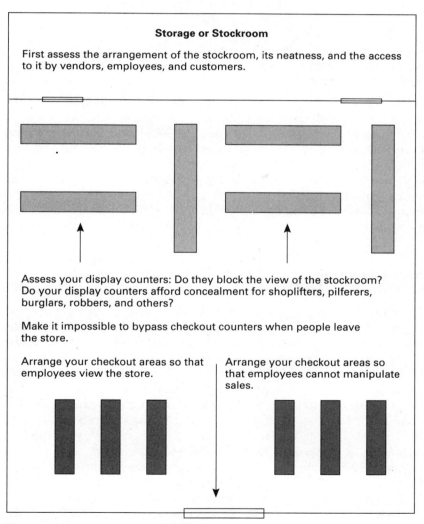

Figure 1-6. Retail store vulnerabilities.

checkout counter is arranged so that the employee sees only the wall or the backs of other employees. Such an arrangement allows employees to steal money or undercharge customers without others noticing.

Display counters also need your attention. Often, retailers become so focused on merchandising and sales that they move the counters close together to display more inventory. However, such an arrangement also heightens opportunities for covert thefts, thus offsetting any profit that might be gained. Finally, stockroom access and neatness need to be controlled to protect business inventories.

Security Technique 7: Look for Vulnerabilities in Your Warehouse Operations

Warehouses often supply opportunities for truckers, employees, burglars, and others to steal merchandise, equipment, and other assets. Most warehouses rely on documents and on shipping and receiving procedures as controls. These do have an important role, but arrangements for holding inventory are equally important. When you begin your assessment, first ensure that your document procedure corresponds with internal controls.

Figure 1-7 shows some key elements in the physical arrangement of a warehouse that add asset control and eliminate opportunities for theft.

Security Technique 8: Check Out Your Manufacturing Plant

What you manufacture has much to do with your assessment of business vulnerability. If your plant makes heavy steal beams, for example, the possibility of anyone stealing them easily diminishes dramatically. But even if your product is relatively theftproof, you need to safeguard administrative supplies, equipment, computer systems and programs, tools, and other items present in the workplace. By contrast, if your plant makes electronic items or specialized costly equipment that is easy to steal and sell, the threats to your business assets increase. In this case, you have to consider the cost of product theft, plus the cost of replacing lost items. Each cuts deeply into business profits.

Intangible assets also need protection. Your manufacturing processes and ingredients, customer lists, bids, expansion plans, and financial records may be highly valuable to a competitor. Manufacturing processes in particular can be lucrative items to steal. Although the pieces of paper in themselves have no tangible value, your business can suffer heavy losses if the information is acquired by a competitor. You

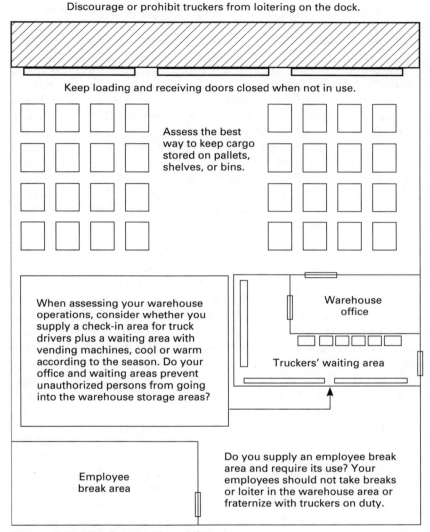

Figure 1-7. Warehouse vulnerabilities.

need to assess your internal protective measures to determine the viability of theft from internal or external sources.

Testing Your Management Controls

After assessing your external and internal asset protection levels, you need to begin looking at management controls. As noted above, information has great value whatever the type or size of your business. Suppose that you operate a chain of retail stores, and each of them prospers. A potential competitor will be highly interested in how you manage to do so well. Maybe your success involves a confidential formula that you discovered after years of trial and error and hard work. Access to the information in your files or computer will enable a competitor to test out the formula and find ways to cut into your hardearned profits. Your assessment effort needs to include a hard look at how valuable your business information is to competitors or potential competitors and how well you safeguard it. If employees have access to that information, you need to develop systems to protect against employee theft, especially from copying documents or computer disks.

You also need to install protective measures for support items used in your business: administrative supplies, tools, containers, and myriad other items. Remember that everything stolen from your business has to be replaced at cost. Thus your loss of a $100 support item really costs you $200. That loss comes directly from "net profit," because the support item, though necessary for your business, is not available for resale.

One of my clients, a major worldwide fast-food chain, asked me to determine why two of its restaurants did record-setting business yet consistently lost money each month. My assessment disclosed regular theft of support items and raw products by vendors and employees. The stores had no daily inventory-taking procedures to detect such a problem. However, after the company established effective controls through a combination of regular inventories and other countermeasures to protect raw products and support items, the losses stopped. Thereafter the two restaurants made record-setting profits.

Interestingly, employees who waste or steal support items rarely view it as a crime. An employee who handles cash might not consider taking even a penny of it, although lax controls offered an opportunity to steal money easily. Yet the same employee might think nothing of taking administrative supplies or other business support items that cost the business hundreds or thousands of dollars in net products. You do not have to believe that employees, vendors, and others will steal you blind (although often they will), but you do need to be aware that many people do not see their theft of supplies as a crime against your business and rarely realize that it can lead to substantial business losses.

Security Technique 9: Audit the Lock and Key System

Perhaps the most basic safeguards of a business are locks and keys, yet these are often the least controlled elements. You need to determine if their control will help you protect your business assets. Check the quality of your locks first. Many generic locks can be opened easily with lock picks, or with makeshift tools such as paper clips. Any exterior or interior door that protects important assets and that cannot be observed readily should have a pick-resistant lock reinforced with a dead bolt. You also need to control the issuance of keys and which employees have access to them.

Remember that your assets include how much you spend on controlling locks and keys. One of my clients had dozens of stores across several states. I found that its lock and key system was not only inefficient but costly. Each store had several locks for exterior and interior doors, each with a separate key. The company employed five managers at each store—one general manager and four assistant managers—and gave each one a set of keys for the entire store. Whenever a manager resigned or was transferred, a local locksmith would rekey all the locks, and the company would issue a new set of keys to all store managers. The company spent about $150,000 each year on this process, believing that it prevented former employees from copying a key and returning to steal assets.

However, my assessment revealed a more efficient method. The company converted to a "core lock" system for one outside door and issued only one key to each manager. The regional operations manager would come to the store after a person left or was transferred and change the core with a master key that came with the system. The manager then would collect all the keys for the old core and put them into stock for use later at another store. The regional manager kept a list of where each core was placed and maintained a rotation schedule. Next, the company "blanked" other access doors so they could be unlocked and locked from the inside only. The general manager maintained one set of keys for the interior locks and personally unlocked any other doors for other employees or managers. Because the new system removed the need for new locks or keys and restricted store access to only one door, security increased and the cost of locks and keys fell to zero.

If your assessment shows that your business issues keys to all locks, with locksmith bills eating into profits, consider decreasing or eliminating the expense with a core system similar to the one described above.

You also need to assess the locks to exit doors, such as those that lead to trash disposal containers, outside stockrooms, and fire exits. These doors should not have locks allowing access from the outside; blank them out instead. Next, determine if the door serves as an emergency or controlled exit. If so, does it have an alarm crash bar to alert you to any exit by employees and others seeking to surreptitiously leave the building with your assets? Crash bar devices shut off only with a key, and that key should stay on your one set of internal masters.

Security Technique 10: Evaluate Your Control of Incoming Packages and Containers

Offices and retail stores experience the greatest amount of loss from thieves who use packages, shopping bags, briefcases, large purses, and other items to conceal assets and carry them out of the business. You need to determine how these containers enter and leave your business. How much a threat this form of theft poses depends on the nature of your business. Will your products or merchandise fit easily into "carryout" containers? Do you protect small items by locking them in glass cases? How closely do you control supplies and other items in your office?

Security Technique 11: Assess Your Control over Vehicles

Allowing employees, vendors, customers, visitors, and others to park near your business can also abet theft. This is especially true in manufacturing, warehousing, and retail operations, which typically have products or merchandise that can be easily hidden in a vehicle. The best way to assess potential losses from this source is to consider how you would steal your own assets and conceal them in a vehicle. Decide if it is possible to control parking arrangements to thwart that threat. Later in the book, I will give you examples of parking security techniques.

Quick Reference Guide to Chapter 1: Assessing Your Threats

- *Remember that you're in business to earn money.* Just as a successful venture needs a comprehensive business plan, so it needs a sound asset protection program.

- *Eliminate the opportunity.* Loss of business assets can happen only when you supply criminal enterprise with the opportunity to commit a crime. Where there is no opportunity, there can be no crime.

- *Losses from your business siphon net profit.* Whenever anyone steals your business assets you lose double the value, because you have to replace them.

- *Do not wait for a problem to develop before creating your asset protection program.* If you do not have a loss problem now, wait a while and chances are you will. You might also not recognize how your loss is happening. Whatever your situation, don't wait for problems to develop. Keep your hard-earned profits by eliminating the opportunity for crime now.

- *Begin effective asset protection management by assessing your business vulnerability.* Do not guess at potential problems. Put yourself in the criminal's place and determine how you would target your business at vulnerable points.

- *Use a systematic process to assess vulnerabilities leading to asset loss.* The only effective assessment is one with a beginning and an end.

- *Work from the outside in.* The best place to begin your assessment of vulnerability is on the exterior of your business.

- *Do not block windows that give passersby, security, and police patrols a view into your business.* You need to weigh the benefits of covering windows with advertising posters against the increased risks of armed robbery and burglary and the resultant business losses and dangers to employees and customers.

- *Continue your assessment on the interior part of your business.* When you have compiled a list of vulnerabilities outside your business location, begin looking inside, especially for physical protection problems.

- *Complete your assessment by considering administrative and support items.* Look carefully at your waste and at losses from the excessive disappearance of support items versus their proper use in sales or services. Also, remember that employees do not always believe they are stealing when they take administrative supplies. Still, the costs are considerable. For example, if each month 500 employees take home a pen that cost you $1.59 plus a pad of lined paper you bought for $1.10, the monthly drain on your net profit is $1345.

- *Consider potential problems from packages, briefcases, purses, and other containers.* Do not overlook problems from container sources, especially if your type of product is vulnerable to walkaway theft.

- *Check to determine if vehicles near your business afford a means of asset theft.* Warehouses and supply or retail businesses are especially vulnerable to thefts in vehicles.

- *List your findings in categories and set priorities for taking corrective measures.* If your vulnerabilities seem overwhelming, you need to prioritize them, dealing with the largest loss potentials first.

2
Choosing the Best Locks for Your Business

Locks are the basic safeguards of a business. However, you should consider them as only the first layer of security. Whatever their quality or cost, locks are at best "delaying" devices, not total bars to entry. Many ingenious locks have been developed over the years, but equally ingenious criminals have found means to open them surreptitiously. Most locks installed during construction of a building, including basic cylinder locks, are vulnerable to picking. A skilled burglar can pick them open in 20 seconds, leaving no traces of entry. Other types of locks require more time and expert manipulation to overcome, but remember that all locks will succumb to force and the proper tools.

Locks originated in the Near East. The oldest known example comes from the ruins of a palace in Khorsabad, near Nineveh in modern Iraq. Possibly 4000 years old, it is remarkably similar to the modern pin tumbler lock. Today countless other types of locks are available, and in the following pages I will help you select the best ones for your business.

Six Common Types of Business-Related Locks

How much protection a well-built lock affords depends on the resistance of the locking mechanism to picking, manipulation, or drilling. The type of lock you choose for your business will be governed by how much security you need and your budget.

Six types of locks are in common use in business: cylinder locks, wafer locks, warded locks, combination locks, core-lock systems, and cypher locks. The first three have basic key-operated mechanisms. They are more popular than other devices because they are generally cheaper and easier to maintain. Details on the mechanisms of several of these locks are supplied in a later section.

Cylinder or Pin Tumbler Locks

The basic cylinder lock is the most commonly used in business. It is also the least expensive, most standardized and easiest to open (see Fig. 2-1). A skilled burglar using lock picks can open a cylinder lock in 20 seconds or less.

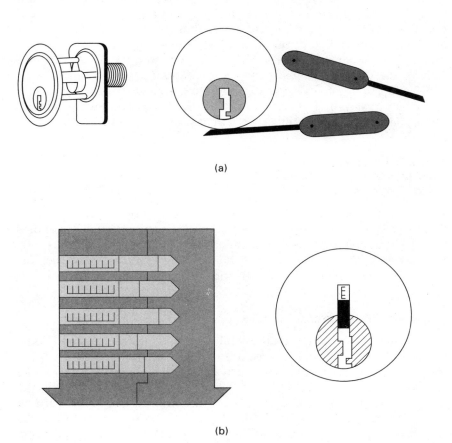

(a)

(b)

Figure 2-1. Basic cylinder lock design. (a) External view. (b) Interior design.

If you use cylinder locks, you are leaving yourself open to lock picking and to the possibility that the keys will be lost or copied. A large problem with business-related cylinder locks involves the routine issuance of keys to authorized persons. The employee receiving a key understandably worries about losing it and discretely makes a duplicate. Or the employee is afraid of forgetting to bring the key to work and makes a copy to keep in the car or tuck in a wallet. Sometimes employees lose the key you gave them, and use their duplicate to prevent you from learning about their carelessness. They then make another duplicate to replace the one they lost.

Issued keys multiply dramatically in this manner, and their existence is a boon to criminals. First of all, illegal entry by key suggests an "inside" job. When a burglar uses a key to enter your business, suspicion of theft normally focuses on an employee instead of on an outsider. Second, criminals can find ways to make an impression of a key to your business, such as using the spare key that an employee leaves in his or her car. Not only can a duplicate key be obtained from an impression of the original, but an experienced lock picker can simply study the impression to determine how to use picks for fast entry into your business.

Selecting the best key locks for protecting your business begins with a clear understanding of types available, the protection they offer, and the best applications for them given their inner workings.

Security Technique 12: Application of the Basic Cylinder Lock

The basic cylinder lock has a brass, steel, or aluminum construction with five pin sets and springs (some later models use six pin sets). These pins and their springs prevent the cylinder from turning. When the proper key is inserted, the pins line up at the top edge of the lock cylinder (the shear line), allowing the key to turn and either open or close the lock.

Wafer or Disk Tumbler Locks

Wafer locks are used with automobiles, desks, file cabinets, and some padlocks (Fig. 2-2). The locks feature several wafers in the core or plug (the part of the lock that turns). The wafers are under spring tension and protrude outside the diameter of the plug into a shell formed by the body of the lock (see Fig. 2-2*b,* locked), keeping the plug from turning and keeping the lock locked. Insertion of the proper key forces the wafers to withdraw from the shell into the diameter of the plug, allowing the plug to turn (see Fig. 2-2*a,* unlocked). A wafer lock in a door or a

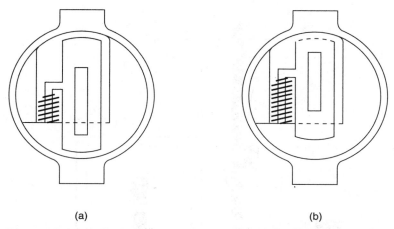

(a)   (b)

Figure 2-2. Wafer lock operation. (*a*) Unlocked—wafer is clear of shell. (*b*) Locked—wafer is locked in shell.

desk has an easily manipulated spring-operated bolt. If used in a padlock with a spring-operated bolt, it is easily "shimmed" open. The manipulator inserts a thin *shim* (spring steel strip) to apply moderate pressure while striking part of the lock sharply with some hard object, hoping to make the wafers bounce out into the diameter of the plug. The pressure applied will cause the plug to turn when the wafers bounce out.

Warded Locks

Warded locks feature obstructions (wards) in the keyway (keyhole) and inside the lock to prevent all except the properly cut key from entering or working the lock. As shown in Fig. 2-3, the key must have the exact ward cuts to bypass the wards. The popularly known *skeleton key* is designed to bypass most wards in any warded lock. However, it is not the only device available. A piece of wire bent in the right shape can bypass the wards and reach the bolt of the lock.

Combination Locks

The popular combination lock works effectively in padlocks, vaults, and door locks. It evolved from the "letter lock" developed in England at the beginning of the seventeenth century. Originally, letter locks were used only for padlocks and trick boxes. In the last half of the nineteenth century, they were also used for safes and strongroom doors and

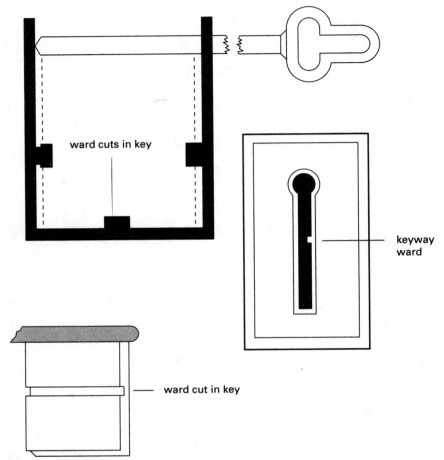

Figure 2-3. Interior wards and ward cuts in keys.

proved to be the most secure protection. Their added security arises from the almost infinite combinations of letters or numbers and the absence of keyholes into which an explosive charge might be placed.

Security Technique 13: Application of Conventional Combination Locks

A combination lock has the same number of tumblers as there are numbers in the combination. Therefore, a lock with three numbers in its combination has three tumblers, one with four numbers has four tumblers, and so on.

A conventional combination lock can be opened surreptitiously by determining the settings of the tumblers through the senses of touch and hearing. It can take several hours to manipulate some combination locks, but a skillful manipulator can open an average conventional combination lock (three-position dial) in a few minutes.

Combination locks have merit, but like key locks they pose potential security problems. Employees may write down the combination in several places to avoid forgetting it. Also, an attentive observer can watch an employee open a combination lock and memorize the numbers for surreptitious entry after business hours.

Manipulation-Resistant Combination Locks. The design of a manipulation-resistant lock prevents the opening lever from touching the tumblers until the correct combination is set. This lock furnishes a high degree of protection for businesses, especially for those that have high-value inventory or sensitive areas that are not under observation. However, manipulation-resistant locks are only as secure as their combination.

Other Combination Locks. Combination locks with four or more tumblers should be considered for most unobserved business environments. Remember that locks are primarily delaying devices and do not provide absolute security. You need to consider how long a delay the lock will provide to allow other security techniques to come into play: security or police patrols, lighting, alert passersby, and other countermeasures.

Relocking Devices. A relocking device on a safe or vault door furnishes an added degree of security against forcible entry. The device appreciably increases the difficulty of opening a combination lock container because to do so requires punching, drilling, or blocking the lock or its parts. This device is recommended for heavy safes and vaults. An intruder needs cumbersome equipment and will make noise and create odors that others may detect. Relocking devices are an excellent means of delaying intruders until other techniques reveal their presence.

Core Locks

The interchangeable core system uses a lock with a core that can be replaced by another core using a different key. Its main advantages are:

- Fast replacement of cores when necessary, instantly changing the matching of locks and keys and minimizing risk to business security
- The keying of all locks into a general locking system
- Economical security through reduced maintenance costs and new lock expense
- Flexibility when engineered to your business needs
- Simplified record keeping

Security Technique 14: Application of Interchangeable Core Locks

Core locks combine excellent security management with long-term cost-benefit advantages. (See the example discussed in Chapter 1.) A core-lock system should be chosen for maximum key duplication and manipulation control. Some lock manufacturers have designed locks with special internal mechanisms that make picking or other techniques difficult and time-consuming. These models also have specially engineered keys that cannot be duplicated by anyone other than the manufacturer. Even making an impression of the key will not help a thief because of the engineering process. Although the initial cost might be more than you planned to spend, a core-lock system will save money and enhance security in the future.

Cypher Locks

A cypher lock is a digital (push buttons numbered from 1 through 9) combination door-locking device used to deny access to anyone not authorized or cleared for entry into a specific business area, such as a stockroom. This type of lock operates on the same principle as a combination lock except that it is more difficult to overcome. Its weaknesses stem primarily from controlling access to the combination.

The Pros and Cons of Lock Security

In the competitive market of security equipment, each manufacturer strives to develop a better product that will supply greater security and generate stronger sales to businesses. However, as noted earlier, a lock is only part of your "layered" system, not the ultimate solution that it is sometimes portrayed to be. Often, a cheap lock gives you the same amount of protection as a costly high-tech lock, depending on your business environment. For example, placing costly locks on doors that can otherwise be forced open does not increase business security. Much of your security management, based on wise, cost-benefit, and protection-enhanced choices, depends on understanding or rethinking the role that locks play in your total security system. The following information provides you with key operating elements you need to make prudent choices.

The Combination Mechanism

The operating principle of most combination locks is a simple one. The operator uses numbers (or other symbols) as reference points to align tumblers so that the locking parts of the lock move to an unlocked position.

Figure 2-4b represents a three-tumbler combination lock mechanism. The letter A represents the dial, which is firmly fixed to the shaft, E. Any movement of the dial is directly imparted to the shaft. Letters B, C, and D identify the tumblers. Each tumbler resembles a disc with a notch cut into its circumference. The technical term for this notch is a *gate*. D represents the driver tumbler. The driver, like the dial, firmly fixes to the shaft so that when the dial moves, the driver tumbler also moves. B and C are rider tumblers. They merely rotate around the shaft. Moving the dial does not immediately also move the rider tumblers.

To operate the lock, you must align the gates with the fence, as shown in Fig. 2-5. When the fence is free to move into the space made by the gates, the lock will operate. Figure 2-6 shows an improper alignment of gates and fence. That happens by applying the wrong combination, which prevents operation of the lock.

To determine possible combinations on a lock, you raise the total number of reference points on the dial to the power that is equal to the number of tumblers. Suppose a lock has 40 numbers on the dial and a three-number combination indicating three tumblers in the lock. Therefore, possible combinations are 40^3, or 64,000. How can someone find one combination out of 64,000 in less than an hour?

Inexpensive combination padlocks have a serial number stamped on the back. Code books (available from locksmith supply houses) list the lock serial number and the combination for that lock. Incidentally, a code for some Master brand combination padlocks is available at low cost from major lock suppliers. This is one easy technique that intruders can use to neutralize your locks.

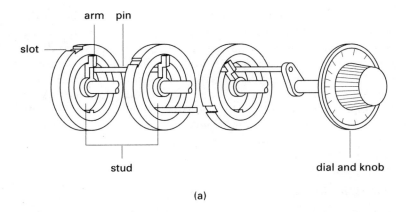

(a)

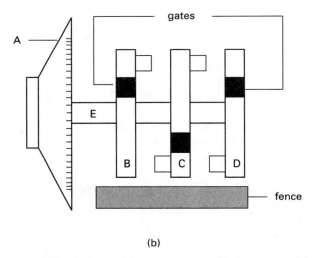

(b)

Figure 2-4. (*a*) Three-tumbler combination lock system. (*b*) Inner mechanism.

A primary weakness of inexpensive locks is the imperfect tolerance between the widths of the gates and the widths of the fence. As a result, an intruder can manipulate the lock open without applying the exact combination. If the combination is 1-3-8, for example, the lock might also open on 2-4-7 or 1-4-9. Therefore, an intruder will try every other combination instead of every combination. This greatly reduces the time required to overcome the lock.

There are still other ways to neutralize small combination padlocks.

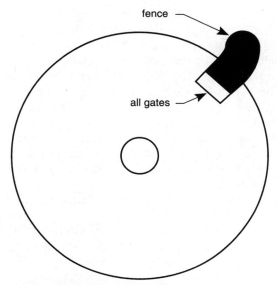

Figure 2-5. Example of combination lock with gates aligned and fence in open position.

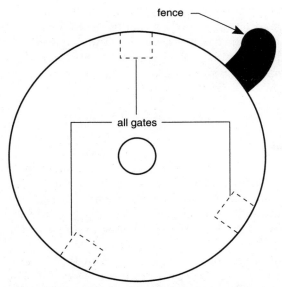

Figure 2-6. Example of combination lock with gates not aligned.

Most models have a spring-loaded bolt (that part engaging the shackle). A properly executed blow to part of the lock will make the bolt disengage from the shackle, and the lock will open. A common term for the technique is *rapping*. The combination padlock can also be opened by shimming the spring-operated bolt with a thin piece of metal known as a *sneaker*. This is a quiet, simple, and fast operation.

A combination safe lock can be manipulated, often with the same success as a simple lock. However, it is not as easy as it appears on television and in the movies. Most big combination locks have very close tolerances between gates and fences, balanced tumblers, and false gates to foil surreptitious burglary attempts. Manipulating such locks takes skill, knowledge, and equipment.

Combination locks and padlocks with changeable combinations are the most cost-effective security measures. However, all good systems need attentive management to sustain their advantages. Lock combinations must be changed under strict control, and accurate records must be kept. You should change combinations on a routine basis, especially each time you change employees. Retain the change keys in the security office and keep a log of changes. For optimum security, change all locks without the authorized user knowing the time or date.

In situations that involve the theft of information by competitors, an unauthorized employee might have obtained the combination to a container with sensitive information. If an employee who is interested in or intends to steal information knows when the lock combination is scheduled to be changed, he or she can copy valuable information on a weekend or at night under the guise of working late and can then deliver or sell the information to a competitor. Unannounced lock combination changes might prevent the loss of information or other valuable items.

The Warded Mechanism

Most warded locks are made of laminated metal, which makes them appear strong and secure. Their mechanical operation is identifiable by a free-turning keyway. A nail file or similar instrument will turn the keyway but not operate the lock. The keyway is simply a guide for the key, not a functional part of the lock. However, when you insert the tool too far into the lock, it will no longer turn the keyway.

Figure 2-7 shows a warded padlock. Characteristics of this type of lock include a shackle secured by a flat spring rather than a bolt. The leaves of the spring compress on the sides of the shackle, and engage a notch on each side of it. To open the lock, you need only spread the

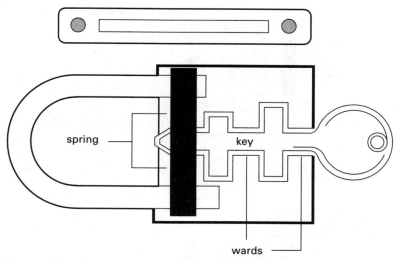

Figure 2-7. Warded padlocks with spring-secured shackle. This is an open view with key inserted.

leaves of the spring—with the proper key, with a specially designed key, or with an ingeniously bent paper clip.

Warded locks and padlocks are common in older buildings and in storage sheds. The locks are primarily for show and offer little security.

In fact, any lock that relies entirely on wards is not good for security purposes. Most modern locks use wards to curtail insertion of unauthorized keys. However, the locks have additional security features besides the wards.

The Pin Tumbler Mechanism

The pin tumbler lock is widely used in commercial security. It generally has greater security than warded or wafer tumbler locks. The pin tumbler lock uses pins moved by a key. It opens along a shear line that is obtained when the key turns the plug (see Fig. 2-8b).

Pin tumbler locks have a variety of applications, including padlocks, door locks, switches, and machinery. When the pin tumbler operates a spring-loaded bolt, the lock is easily overcome by rapping or shimming it open. Shimming a door lock is also known as *loiding* (from the word "celluloid," a material commonly used in this technique).

As with wafer locks, adding a dead bolt to a pin tumbler lock prevents rapping or shimming. The plug of the lock must turn to operate the dead bolt. Within a dead bolt, the lock must be picked to be opened

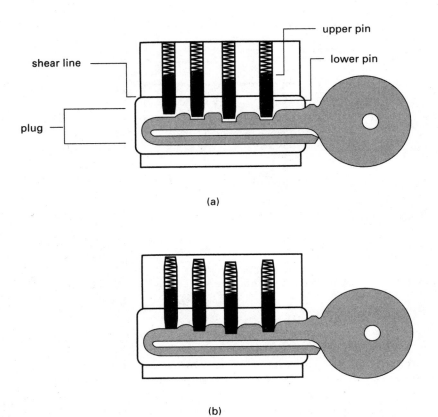

(a)

(b)

Figure 2-8. Pin tumbler lock operation. (*a*) Wrong key. (*b*) Correct key.

or overcome. Some locks feature a secondary deadlocking latch bolt (see Fig. 2-9). Properly adjusted so that when the door is closed the bolt is fully extended into the strike (the recess for the bolt in the door frame) and the secondary bolt is fully depressed, the secondary dead-locking latch bolt prevents loiding or shimming. This kind of lock offers low-cost security for office doors of nonsensitive business areas.

The pin tumbler mechanism (except for one or two makes of pin tumbler locks) lends itself to the technique of mastering. In mastering, the pins are segmented by splits that allow several possible pin align-ments at the lock's shear line (see Fig. 2-10). Thus several differently cut keys can operate the same lock. Because of this feature, mastering makes picking easier for the intruder.

A mushroom or spool tumbler pin placed in a mastered lock makes picking more difficult. Picking tools tend to cant the tumbler sideways and bind it at the shear line (see Fig. 2-11). The mushroom-type pin can also be used effectively on nonmastered locks.

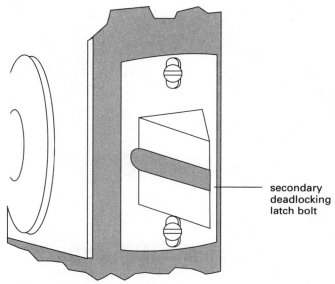

Figure 2-9. Secondary deadlocking latch bolt.

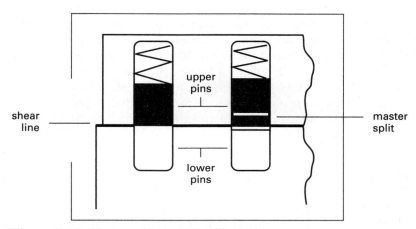

Figure 2-10. Master split in pin tumbler lock.

The Art and Science of Lock Picking

People create locks; other people defeat them. So it is foolish to claim that a lock is pick-proof. At best a lock is pick-resistant, but this is a relative term. In reality, picking locks is based on a few simple principles, including the construction characteristics of the lock. Anyone can

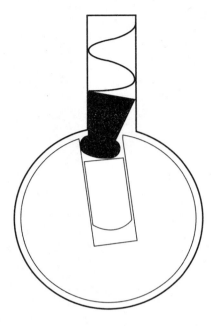

Figure 2-11. Mushroom tumbler action during picking.

purchase lock picks from locksmith supply houses or through a variety of mail-order catalogs. Lock picks can also be created at home or at work, with a few easily found materials. Many locks will open quickly using ordinary paper clips, and can be relocked the same way. Such locks include those found on most file cabinets and office desks.

Figure 2-12 shows a professional lock-picking kit. The two essential picking tools are the tension wrench and the pick. The tension wrench

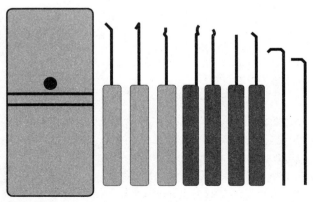

Figure 2-12. Professional lock picking set.

applies light pressure in a rotary motion to the key plug of the lock to aid in finding the bindings or locking tumblers. The wrench holds the tumblers in place as the pick moves the tumblers, one at a time, to the shear line. When all tumblers become aligned properly with the shear line, the lock opens. Closing the lock with picks involves the same process. Picking a lock takes practice, skill, and a little luck.

Despite the ease in picking most locks, intruders prefer other methods. Traditionally, they cut locks, pull them apart, blast them with light explosive charges, or pry them off doors. Sometimes an intruder simply spreads the door frame away from the door to release the bolt from the strike. You can discourage that action by installing locks with at least 1-in bolts and by using grouting to hold the door frame rigidly.

Sometimes, intruders cut the bolts of padlocks or door locks by placing a saw blade in the space between the door and the frame, or by sawing the padlock shank. Good-quality padlocks and door locks have hardened steel in the middle of the bolts to prevent the saw from getting a bite. Intruders may breach the padlock hasp if the screws holding it to the door have not been properly installed.

Security Technique 15:
Application of Dead Bolt Latches

The dead bolt latch can be used on almost any door lock. It is easy and inexpensive to install and increases security. To be most effective, the bolt of the door latch must slide into the door casing from a keeper that is firmly attached to the frame (not the door facing). Look at the examples of the best dead bolt mechanisms in Fig. 2-13. The dead bolt latch is an effective security measure in your business asset protection program.

Controlling Locks and Keys

Controlling locks and keys (or combinations) is of primary importance in safeguarding property and business information and in preventing other criminal activity in your business environment. It is one more layer in your total security system. The following steps are essential. You can easily adapt them to meet your specific security requirements.

Security Technique 16:
Application of Lock and Key Control

For effective lock and key control, you will need accurate records. Also, you need to make periodic physical inspections and inventories to ensure the accuracy of your system.

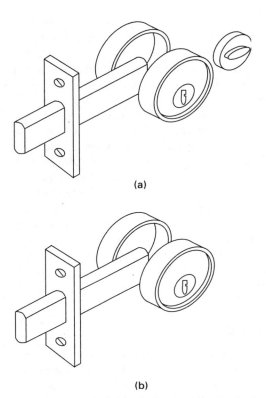

(a)

(b)

Figure 2-13. Double cylinder deadlocks.
(*a*) Bolt may be operated from inside by the
thumb turn or from the outside with a key.
(*b*) Bolt must be operated from either side
by a key.

1. Combinations or keys to locks should be accessible only to authorized personnel.

2. Combinations to business safes and to padlocks securing sensitive
   business information need to be changed at least once a year. Other
   types of locks need regular rotation or core changes. Circumstances
   might call for changing combinations before scheduled dates. The
   following situations prompt immediate changes to combination or
   other types of locks:

   ▪ Loss or possible compromise of the combination or key
   ▪ Discharge, suspension, or reassignment of any person who
     knows the combination or was assigned a key
   ▪ Receipt of a container with a built-in combination lock

3. Key padlocks should be rotated frequently.

4. When choosing combination lock numbers, avoid multiples and simple ascending or descending arithmetic series.

5. When padlocks with fixed combinations include bar locks as supplemental locking devices, keep an adequate supply so locks can be interchanged frequently. Never use fixed-combination locks to protect sensitive business information.

6. Protect your records with highest degree of security as the most sensitive material they contain.

7. Limit the number of keys issued. Access to keys needs your constant supervision.

8. Store your business keys in a locked, fireproof container when they are not in use.

9. Make access lists for people authorized to draw keys, especially when you delegate responsibility to managers or other assistants.

10. Do not allow employees to keep keys or take them off the business premises.

11. In a manufacturing or similar business operation, make sure your supervisors inspect all key containers after each shift; all keys must be accounted for. Keep a log for supervisors to sign.

12. Maintain key control records on all key systems. Accountability concepts include key cards and key control registers. Each record must include at least the following information:
    - Total number of keys and blanks in the system
    - Total number of keys by each keyway code
    - Number of keys issued
    - Number of keys available.
    - Number of blanks in stock for each keyway code
    - Names of people issued keys

Designing a Key Control System

You need to inventory your business key systems regularly. Requests from employees for issuance of new, duplicate, or replacement keys need your personal attention. Plan to provide maximum security with a minimum interference to your business operations.

Basic requirements for designing effective key control systems are as follows:

1. Use high-security pin tumbler cylinder locks in sensitive business areas, such as research and development, storage areas for high-value items, information libraries, and computer sections.

2. Develop key control systems to prevent usable keys from being left in possession of contractors, vendors, or other unauthorized people. Use locks with restricted keyways and issue new keys on blank stock that is not readily available to local commercial key makers.

3. Prevent master keying except in those business areas you regularly observe and areas that contain minimum-value items or items that are difficult to carry away. Using several shorter pins to master a pin tumbler lock is less secure than using a maximum number of pins on a lock that is not mastered. Mushroom-type pins should be added to each lock.

4. All keys and locks (or lock cylinders when appropriate) in a master-keyed system need to be given an unobvious number system. Imprint the words "Property of (Business Name)—Do Not Copy" on all master and high-level security control keys.

Security Technique 17: Conduct a Vulnerability Survey

Since all businesses have unique conditions and requirements, their key control systems will vary. Before you establish a lock and key control system, conduct a vulnerability survey to determine your actual requirements. You will need to identify all warehouses, shops, storage areas, safes, filing cabinets, and other locations that need the added protection of locks and the security of keys. When you make that decision, include it in your asset protection system plan, along with the following information:

- Location of key depositories (when applicable)
- Keys (by building, area, or cabinet number) turned into each depository.
- Method of marking or tagging keys for ready identification
- Method of controlling the issuance and receipt of keys to employees, including a registry of persons authorized to possess business keys
- Action called for when keys are lost, stolen, or misplaced.
- Frequency and method of lock rotation
- Assignment of responsibility for keys, by job or position title and department
- Emergency keys, readily available
- Other controls as thought necessary

Security Technique 18: Choose Keys and Locks for Sensitive Information and Property

Some businesses have special areas that require added security measures—for example, a computer center where access is controlled and information is stored. Whatever the reason, added controls—ranging from vaults to special storage rooms—means layered security and high-security locks that make entry difficult for the determined intruder and impossible for the casual pilferer.

All doors used for access to sensitive-item storage rooms must be locked with high-security devices. The most secure door should have a high-security padlock and hasp. I recommend a steel door with two ¼-in or thicker steel bars running across the door. The bars can be separate, with a slot at each end slipping over a heavy hasp that is welded to the reinforced steel door casing. Or the bars may have their own welded hinges and be slipped over a welded hasp on only one side. The secondary padlock needs to be of good quality, preferably also a high-security lock. I suggest that the primary lock be a high-security key-operated padlock, with the secondary lock being a high-security changeable combination lock. That arrangement can baffle the most determined intruder.

Added protection comes from a double-door or triple-door system, depending on the importance of the room, building, or area you are protecting. In a double-door system, the first door or layer should have the same protective measures as those described above. The second door should have a single bar secured with a high-security lock. In triple-door arrangements, you need only add a third door secured in the same manner as the second door. You might also place a high-security cylinder lock on the handle of the door as a supplement to the other padlocks.

Figures 2-14 and 2-15 show the single-, double-, and triple-door systems. Your high-security locks must meet the following specifications, depending on their application:

- A key-operated mortise or rim-mounted dead bolt for door handle security.
- A dead bolt throw of 1 in
- A double-cylinder design
- A cylinder with five-pin tumblers, two of which include mushroom or spool-type drive pins
- At least 10,000 key changes
- No master keying of locks
- A visible bolt made of steel or with hardened, saw-resistant inserts

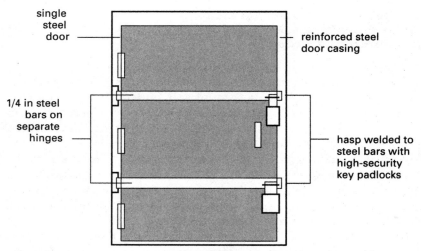

Figure 2-14. High-security storage—single door with double locks. Note: High-security changeable combination locks are also acceptable.

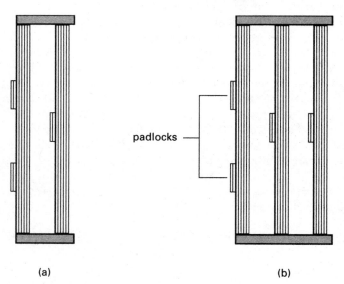

(a)   (b)

Figure 2-15. Outer and inner steel doors with (a) double and (b) single locks.

■ Padlocks with a high-security rating

■ High-security combination locks with changeable combinations

Vehicles and storage facilities in which items are stored must be secured by approved secondary padlocks. Doors that cannot be secured

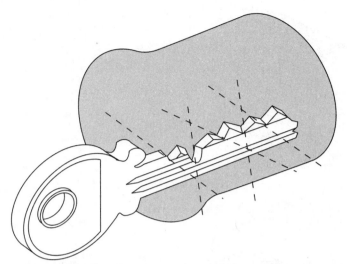

Figure 2-16. High-security key for cylinder or padlock. The key biting angles cannot be duplicated on ordinary key-cutting machines.

from the inside with locking bars or dead bolts need secondary padlocks. Figures 2-16 through 2-19 show high-security door cylinder locks. High-security key padlocks operate in the same manner.

Security Technique 19: Protect Facilities Outside Your Control

When a specific business facility is not directly under your control, special arrangements are required for high-security keys and locks. Key control registers need to contain the signature of each person receiving the key

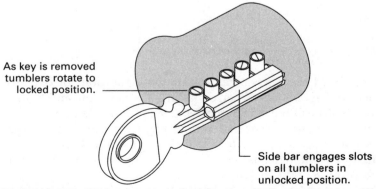

As key is removed tumblers rotate to locked position.

Side bar engages slots on all tumblers in unlocked position.

Figure 2-17. High-security cylinder or padlock mechanism. All tumblers must be at specific heights and angles for unlocking.

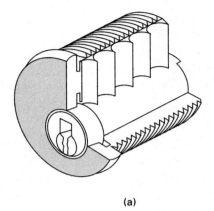

(a)

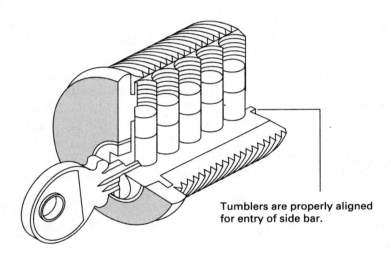

Tumblers are properly aligned
for entry of side bar.

(b)

Figure 2-18. High-security cylinder or padlock interior.

from the appointed manager, including the date and hour of issuance,
serial number of key, initials of person issuing the key, date and hour of
return, and signature of the person receiving the returned key.

Make sure you provide clear instructions and procedures. Emphasize
that high-security padlocks are to remain locked to the staple or hasp
after an area or container has been opened, in order to prevent theft,
loss, or substitution of the lock.

Combinations to locks on vault doors or other containers should be
changed semiannually, and the combinations must be safeguarded.

Figure 2-19. High-security lock mechanism with staggered multiple pins.

Padlocks on entrances to sensitive storage facilities need to be rotated at least semiannually, and a record should be made of the date of rotation. Make sure the same locks are not used within an area where the doors or containers protect sensitive items or within the same storage facility. A sound preventive measure is to exchange locks among other areas of the business.

To facilitate fast replacement and regular rotation, maintain available backup sets of locks for at least 15 percent of the business locks in use. Keys to locks taken temporarily out of service must be inventoried and checked when rotated. The loss of or inability to account for any key to a lock makes that lock ineffective for securing sensitive items. Locks or combinations need to be changed in the following circumstances:

- When placed in use after purchasing
- At least semiannually after purchase
- On the transfer, reassignment, resignation, or firing of any person having the combination
- When the combination is compromised or the lock is found unlocked and unattended

Security Technique 20: Develop an Inspection Procedure for Defective Locks

In your security office, keys should be kept in a locked container built of at least 20-gauge steel or material of an equivalent strength. Never leave keys unattended or unsecured. Keys to sensitive buildings, rooms, and other areas need a second container to prevent mistaken issue to unauthorized personnel or removal from the business area. Avoid master key systems in high-security areas. When high-security keys are lost, misplaced, or stolen, you must make sure the affected locks or lock cores are changed immediately. Protect all replaced items against access by unauthorized personnel.

You or your designated manager should conduct a periodic inspection of all business locks to determine their serviceability and condition. You should examine the locking mechanism's effectiveness, look for signs of tampering, and make any necessary replacements. You can test the locks by inserting a test key (any comparable key other than the correct key) no more than $\frac{1}{4}$ in into the keyway. Turn the test key carefully, using the normal amount of force needed to open the lock. If the lock opens during the test inspection, replace it immediately. Take care during inspection to prevent jamming the test key into the key recess. You may be unable to remove the test key and may cause severe damage to locking levers.

Besides regular inspections, you need to institute a program of periodic preventive maintenance. Make sure business locks have adequate lubrication, rust-preventive outer coatings, and clean keyways.

Quick Reference Guide to Chapter 2: Installing a Lock and Key System

- Are all business entrances equipped with secure locks?

- Are entrances always locked when not in active use?

- Are hinge pins to all locked entrance doors spot-welded or bradded?

- Have you appointed a manager to control business locks and keys?

- Are locks and keys to all buildings and entrances supervised and controlled?

- Have you established a clear policy for issuance and replacement of locks and keys within your business?

- Are keys issued only to authorized employees?

- Do you have a clear policy about removing keys from the business premises?

- Are backup keys properly secured in a locked cabinet?

- Do you have records showing the buildings and entrances for which keys are issued?

- Are there records showing the location and number of master keys?
- Are records maintained on the location and number of duplicate keys?
- Are records maintained on the issuance and return of keys?
- Are records maintained on the location of locks and keys held in reserve?
- Is there a key control policy?
- Are locks changed immediately upon loss or theft of keys?
- Are lock and key inventory inspections conducted at least once each year?
- If you use master keys, are they without identifying marks?
- Are losses or thefts of keys promptly investigated?
- Do you personally approve all requests for reproduction or duplication of keys?
- Are locks on inactive gates and storage facilities under seal?
- Do you check sealed locks periodically?
- Are locks rotated within your business at least semiannually?
- Is the manufacturer's serial number on combination locks obliterated?
- Are there measures to prevent the unauthorized removal of locks on open cabinets, gates, or buildings?
- Are all doors locked at the close of the business day?

3
Increasing Structural Security

Increasing structural security begins with the vulnerability assessment recommended in Chap. 1. In that assessment, you identified the weaknesses in your business structure. You are now ready to learn some techniques to strengthen your asset protection program. Remember that in this chapter, as in all chapters, I will supply the maximum security effort. You need to adapt the techniques according to the specific needs and location of your business. One caveat: Do not forget local fire escape laws and common sense. In increasing structural security you must never block emergency exits.

Hardening the Entry Points of Your Business

Statistics show that doors and windows are the most frequent points of attack in a burglary. However, *any* opening is a potential entry point and should receive adequate protection, or "hardening."

Security Technique 21: Harden Doors

Door Frames. You need to ensure that all door frames are at least 2-in-thick wood or metal with a rabbeted jamb, or hollow metal with a rabbeted jamb filled with a solid material that can withstand spreading. When the materials in the door and frame are inadequate, a security gap occurs. An experienced intruder can gain entry in seconds by placing a heavy pry bar between the door and the door frame just below the lock. A quick push releases the lock, and the door opens.

Entry/Exit Doors. All doors in your security system should be 2-in-thick solid wood. If the wood is $1\frac{3}{8}$ in or less, it should be covered with 16-gauge sheet steel or 24-gauge steel bonded to a kiln-dried wood core. There should be a minimum 60-watt illumination above doors.

Door Locks. Each lock should have a dead bolt or dead latch with at least a 1-in throw, plus an antiwrenching collar or a secondary dead bolt with a 1-in throw. As a minimum, use a $\frac{1}{2}$-in throw for low-priority areas. Sliding doors, garage, and loading dock doors need to be secured with the best possible hardware to make opening by intruders difficult.

Security Technique 22: Harden Windows

Windows. Large windows need burglary-resistant material. Smaller windows—such as those in warehouses, shops, and storage buildings—should be painted over when practical, and protected by $\frac{1}{2}$-in round steel bars (or 1-in by $\frac{1}{4}$-in steel flat bars) placed 5 in apart. The bars need to be secured in 3 in of masonry or by $\frac{1}{8}$-in steel wire mesh, with grid spaces no larger than 2 in, bolted over the window. To harden nonventilation windows, glass brick is a secure choice. For windows less than 18 ft above ground level or within 40 in of an interior door handle, use inset lag bolts to prevent intruders from lifting the windows out of their frames.

Window Locks. Use clam shell (crescent) thumbscrew pin-in-hole locks or other locks appropriate to the type of window.

Security Technique 23: Harden Other Openings

Any opening larger than 96 in^2 should be covered with steel bars or a mesh screen to prevent intruders and animals from entering. Critical points of entry are skylights, elevators, hatchways, air ducts, vents, and transoms.

Skylights. Cover skylights with round metal bars, grills, or mesh. Bars should be no less than $\frac{3}{4}$ in in diameter and no more than 5 in apart. Mesh should be at least $\frac{1}{8}$ in thick with grid spaces no larger than 2 in.

To prevent removal of skylights from the outside, secure them firmly with machined round head bolts and use special burglary-resistant glass.

Elevators.   When feasible, install an elevator operator, CCTV (closed-circuit television), or continuous open-listening devices connected to a security control station. As an alternative, use special keys or cards that will limit access to authorized personnel.

Hatchways.   Cover hatchways with 16-gauge steel screwed to wood, secured by slide bars or bolts from the inside or by a padlock.

Air Ducts, Vents, and Transoms.   Ventings more than 8 by 12 in wide on the roof or rear of the building should be protected by round or flat iron or steel bars, secured by nonremovable bolts.

Doors and Windows.   Figures 3-1 through 3-3 provide solutions for increasing the security of doors and windows.

Figure 3-1*a* shows a typical office door. Steel surface plates, a strong frame, and a pick-resistant lock (Fig. 3-1*b*) can decrease its vulnerability. Figure 3-2 shows a typical window construction and Fig. 3-3 the measures that can harden windows.

Security Technique 24: Solve the Crash Door Problem

In the business workplace federal, state, and local safety laws need to be considered. Because emergency exits may create a security problem, they must have crash bar doors that allow immediate and easy exit—even in a smoke-filled corridor or in darkness—but that deny entry from the outside.

Whenever employees or others use a door for exit, security problems are created. Often crash bars can be locked only by pulling them to latch from the inside. When an employee exits through this type of door, it may not relock, creating a breach in the security system. Because this type of door also allows people to depart unnoticed, it can invite criminal activity. Further, an employee could open a crash bar door deliberately, allowing an unauthorized person to enter the area with criminal intent.

Figure 3-4 shows a crash bar door viewed from inside and outside. The door should have reinforced-steel construction whenever possible, and should leave no opportunity for access from the outside. It is a door reserved for emergency exits only.

You can prevent intentional nonemergency use of crash doors with a self-contained alarm that activates when the crash bar is depressed.

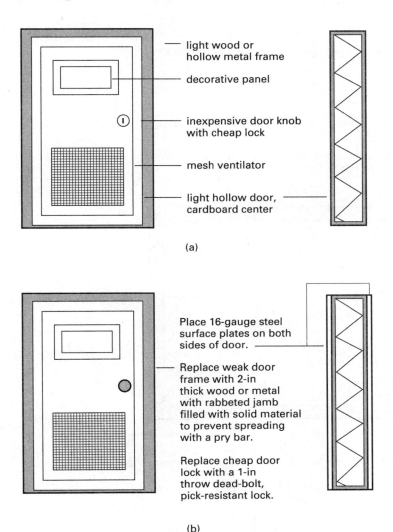

light wood or
hollow metal frame

decorative panel

inexpensive door knob
with cheap lock

mesh ventilator

light hollow door,
cardboard center

(a)

Place 16-gauge steel
surface plates on both
sides of door.

Replace weak door
frame with 2-in
thick wood or metal
with rabbeted jamb
filled with solid material
to prevent spreading
with a pry bar.

Replace cheap door
lock with a 1-in
throw dead-bolt,
pick-resistant lock.

(b)

Figure 3-1. (*a*) Typical office door. (*b*) Viable solution to increase security.

Economical kits are available to install on existing crash bar doors. They have a strong steel alarm box that attaches to the door. The box contains a loud bell or other noise powered by ordinary flashlight batteries. The alarm box is secured with a special key, normally with a round, pick-resistant lock to prevent intruders from disarming the alarm surreptitiously. A paddle extends tightly behind and below the crash bar so that when it is depressed, the alarm sounds until it is reset by key and until the door is relocked.

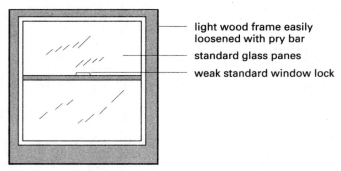

light wood frame easily
loosened with pry bar

standard glass panes

weak standard window lock

Figure 3-2. Typical window construction found in shops, warehouses, storage facilities, and some offices.

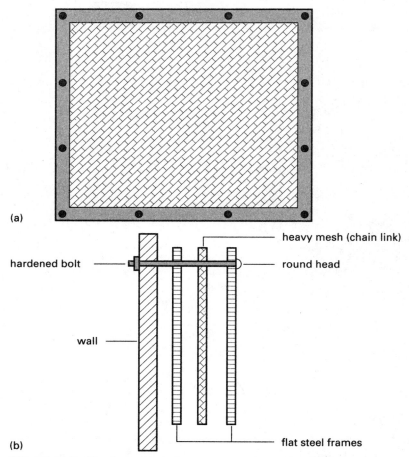

(a)

heavy mesh (chain link)

hardened bolt

round head

wall

flat steel frames

(b)

Figure 3-3. Hardening windows on warehouses, shops, and other areas. (*a*) This solution increases security while allowing ventilation. (*b*) Full window covering.

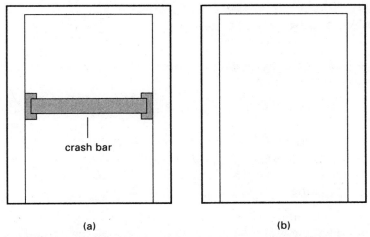

Figure 3-4. Emergency exit crash door. The crash bar attaches to the locking device, which releases when the bar is depressed. (*a*) Inside view. (*b*) Outside view.

Place a warning sign on the door that the alarm will sound when the crash bar is depressed. The sign will discourage employees from using the door as an exit unless a true emergency exists, and will deter unauthorized use for criminal purposes. Figure 3-5 shows the crash bar alarm system. You can purchase the entire assembly with the alarm built in when establishing new installations.

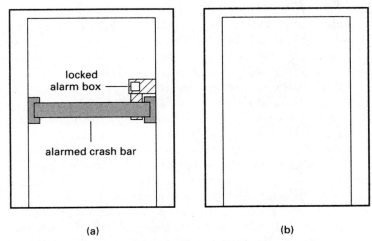

Figure 3-5. Viable solution to crash bar doors serving as emergency exits. (*a*) Interior view. (*b*) Exterior view.

Hardening Exterior Walls and Roofs on Your Business Premises

Two main approaches to hardening exterior walls and roofs increase your asset protection program:

1. Harden walls (and roofs) during construction of a new building or remodeling of an old one. This is practical when you own your business structure as opposed to when you lease the premises.

2. Use some creative thinking to harden existing walls and roofs. For example, increase the strength of walls with heavy fixtures and decorative metal plating.

Remember to think of the roof of your business premises as an exterior wall, with the same vulnerabilities. Do not overlook the possibility of a burglar cutting through the roof to gain access to your business. The roof often provides the best access because it can offer easy concealment and is the least protected part of the structure.

The techniques described below involve exterior walls vulnerable to burglary. To harden interior walls effectively, use the high-tech devices discussed and illustrated in Chap. 4.

Security Technique 25: Harden Exterior Walls and Roofs During Construction

Wood frame buildings are extremely vulnerable to burglary, because wood can be easily sawed or chopped (see Fig. 3-6). In many areas, corporate or industrial buildings have a metal frame and thin sheet aluminum or steel. These structures are popular because of their versatility. Businesses can obtain large square footage space quickly and with less expense by constructing metal buildings. Some of these structures serve as offices, but a majority are used for storage and manufacturing facilities.

As noted in Chap. 1, brick walls are also vulnerable to burglary, even though they may appear invincible. When your assessment of security determines that a building housing sensitive or valuable property is vulnerable through a brick or stone wall, you will need cost-effective solutions. Figures 3-7 through 3-10 present some examples.

Whatever type of security challenges your company faces, the following guidelines will help you overcome the problem of hardening vulnerable walls and roofs quickly and economically. Remember that the same processes of reinforcing walls during construction or remodeling can work effectively on the roof of your business premises.

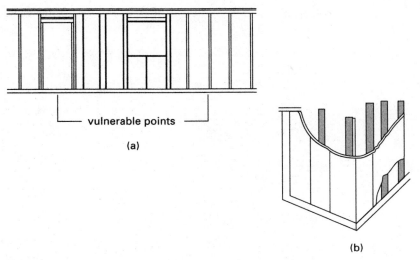

(a)

vulnerable points

(b)

Figure 3-6. Typical frame construction.

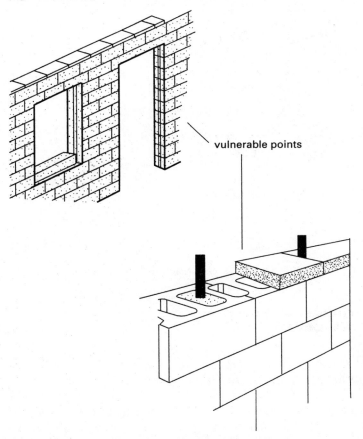

vulnerable points

Figure 3-7. Standard concrete block structure walls.

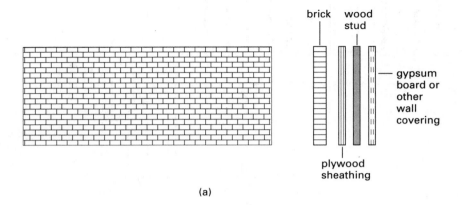

(a)

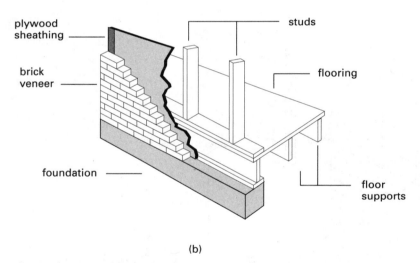

(b)

Figure 3-8. Standard brick structure construction. (*a*) Structure wall. (*b*) Cross section.

Security Technique 26: Harden Existing Walls and Roofs Creatively

If you cannot strengthen your exterior walls and roof with the construction techniques illustrated earlier, consider a few creative solutions using display cases backed with metal barriers. Figure 3-11 shows such an arrangement. In an office, warehouse, or other business environment, you can adapt the same general techniques to increase

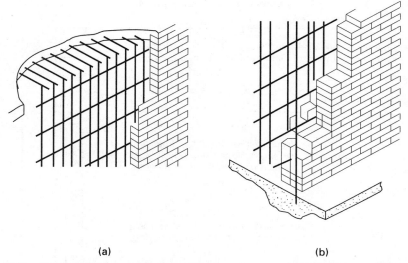

(a)   (b)

Figure 3-9. Reinforced brick structure wall construction. (*a*) Top-of-wall reinforcement. (*b*) Double-reinforced wall.

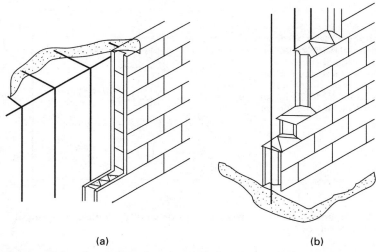

(a)   (b)

Figure 3-10. Reinforced concrete block wall construction. (*a*) Top reinforced section. (*b*) An 8-in concrete masonry wall reinforced with vertical bars and horizontal cross-rod trussed wire.

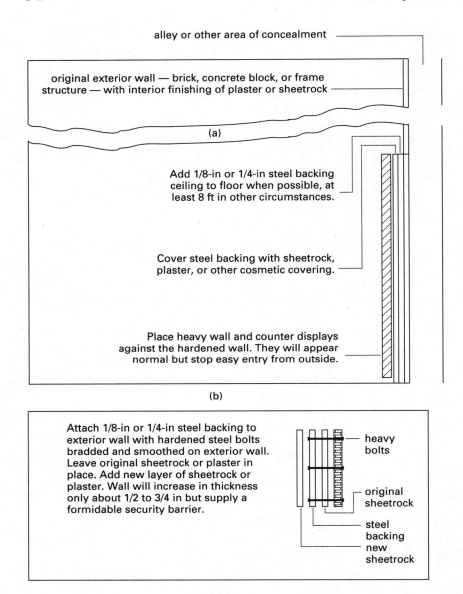

alley or other area of concealment

original exterior wall — brick, concrete block, or frame
structure — with interior finishing of plaster or sheetrock

(a)

Add 1/8-in or 1/4-in steel backing
ceiling to floor when possible, at
least 8 ft in other circumstances.

Cover steel backing with sheetrock,
plaster, or other cosmetic covering.

Place heavy wall and counter displays
against the hardened wall. They will appear
normal but stop easy entry from outside.

(b)

Attach 1/8-in or 1/4-in steel backing to
exterior wall with hardened steel bolts
bradded and smoothed on exterior wall.
Leave original sheetrock or plaster in
place. Add new layer of sheetrock or
plaster. Wall will increase in thickness
only about 1/2 to 3/4 in but supply a
formidable security barrier.

heavy
bolts

original
sheetrock

steel
backing

new
sheetrock

(c)

Figure 3-11. Expanded views of creative exterior wall hardening for retail or
similar business operations.

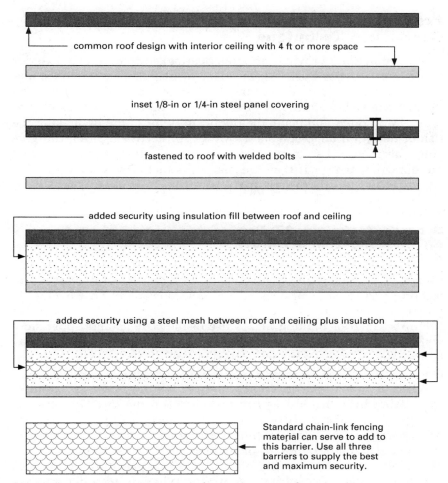

Figure 3-12. Creative techniques for hardening roofs and ceilings.

your security. You also need to strengthen the roof and ceiling of your business. Figure 3-12 shows some creative techniques.

Remember that hardening exterior walls and roofs has only a "delaying" effect, as discussed in relation to locks in Chap. 2. Your goal will always be to eliminate the opportunity to commit a crime against your business. Often that happens by making the risk of entering your business too high or by making the task too difficult. In such a situation, criminal opportunity decreases, the risk of detection increases, and the price becomes too high for the burglar to pay.

Protective Lighting: The Overlooked Technique

Protective lighting provides an often overlooked technique of exterior hardening for your business. Burglars, armed robbers, vandals, and other criminals rely on concealment, especially darkness, to carry on their activities. Effective exterior and interior lighting work together to reduce, and often eliminate, the opportunity for crime by making the risk of detection unacceptable. However, protective lighting can give you a false sense of security, especially if the lighting of power sources are not protected, or if they are placed in such a way as to provide concealment areas for the criminal.

Security Technique 27: Use Effective Lighting to Harden Exteriors

Protective lighting is an essential element of your integrated asset protection program and supplies an added dimension of "hardening" to your external structural security.

Indirect interior lighting makes use of the ceilings and upper sidewalls of your business for redirecting and diffusing light given by lamps. This is in part a matter of the light reflecting properties of various colored surfaces. The same is true of your external lighting.

When planning your exterior protective lighting, give first priority to creating high brightness contrast between the intruder and the background. Dark or dirty surfaces require more light to produce brightness than do clean concrete, light brick, or grassy areas. When the same amount of light falls on an object and on its background, the ability of a passerby, security officer, or police patrol to see the area depends on the reflected light contrasts. When an intruder appears darker than his or her background, an observer sees primarily the outline or silhouette. You can deter burglars or other intruders who depend on dark clothing and even darkened face and hands by using light finishes on the lower part of your business building or structure.

Security Technique 28: Apply Systems of Exterior Protective Lighting

Two basic systems, or a combination of both, will serve to provide you with a practical and effective lighting program:

1. Supply light to boundaries and approaches to your business premises.

2. Light the area and structure within the general boundaries of your
   business location.

The goal of these techniques is to discourage or deter forced entry
into your business and to make detection likely should the intruder
even attempt. Proper illumination may lead a potential intruder to
believe that detection is inevitable.

Types of Protective Lighting Systems

The object of your lighting program is to obtain the illumination necessary
to protect the exterior of your business and its assets. Several types of
lamps, each with various properties, are used in protective lighting systems. For example, some lamps create glare, while others produce a softer,
diffused light. Each is appropriate to different business circumstances.

Incandescent Lamps

Incandescent lamps (glass light bulbs) produce light by the resistance
of a filament to an electric current. Special-purpose incandescent
lamps are manufactured with interior coatings to reflect the light, and
have built-in lenses to direct or diffuse the light. The naked bulb may
be enclosed in a shade or fixture to accomplish a similar result.

Gaseous Discharge Lamps

Gaseous discharge lamps come in three primary forms.
   Metal halide lamps emit a harsh yellow light. They use sodium, thal-
lium, indium, and mercury to conduct current.
   Mercury vapor lamps emit a blue-green light caused by an electric
current passing through a tube of conducting luminous gas. They are
more efficient than incandescent lamps of comparable wattage and
have wider use for interior and exterior lighting.
   Sodium vapor lamps are constructed on the same general principle as
mercury vapor lamps. They are more efficient than the other two types and
are generally used to light large areas with a golden yellow glow. They are
commonly used in parking lots and on roads, streets, and bridges.

Fluorescent Lamps

Fluorescent lamps have large, elongated bulbs that provide a high light
output and have a light life of about 7500 hours. They have a higher
initial cost than incandescent lamps but a lower operating cost.

*Security Technique 29: Protect Power Sources of Illumination*_____

Normally your power source for interior and exterior lighting comes from the local utility company. Certainly, the most secure power lines run underground. In addition, many businesses have lines running from an outdoor pole. Few burglars will consider trying to disconnect or sever such lines, because doing so is likely to cause a power outage over a broad area. However, a skilled burglar will try to disconnect or sever power running from your business to the exterior and sometimes interior lighting. To disconnect all exterior lights would draw attention, so the intruder will disarm only one or two lights, thereby creating shadowed areas of concealment.

*Security Technique 30: Protect Your Wiring Systems*_____

Another important layer in your protective lighting program is the safeguarding of all wires from tampering by an intruder. Your best approach is to install lighting on both multiple and series circuits, depending on the type of illumination used and other design features of your system. Each circuit should ensure that failure of any one lamp will not knock out the others, leaving your business in darkness. Make sure, too, that normal interruptions caused by power outages, overloads, industrial accidents, and building fires will not interrupt your protective lighting capability.

Whenever possible, have power feeder lines underground or tucked inside a secure area of the business. Install emergency lighting with sensors so that if your primary system fails for any reason, the backup system will turn on automatically. (More on emergency lighting is presented later in the chapter.) In all systems, provide for simplicity and economy of maintenance, and schedule regular maintenance and replacement, keyed to the expected life of each lighting element.

You can best protect your lighting power sources by installing a loud alarm (such as a bell) that will be heard throughout the area whenever a wire leading to your protective lighting is cut. The alarm control should be within your business premises, and the alarm should be placed out of sight on the roof or in some other location where it is not obvious or easily accessible. More elaborate systems not only sound an alarm but also alert a central security station, which will send a patrol or notify the police. Such systems are available through security supply houses or local alarm agencies.

The "Best" Lighting for Protecting Your Business Premises

The best types of lighting for protecting a business depend on the type of business and its location. Some or all of the following may apply to your operation.

Continuous Lighting (Stationary Luminary)

Continuous lighting is the most common protective system. It has a series of fixed luminaries arranged to flood a given area continuously with overlapping cones of light during the hours of darkness. Two primary continuous-lighting techniques are glare projection and controlled lighting.

Glare Projection Lighting.   The projection technique works best when the glare of lights directed across an area will not become annoying to those using the area. Glare supplies a strong deterrent to crime by making it difficult for potential intruders to see and impossible for them to conceal either their presence or their activities. Figure 3-13 shows a combined glare and continuous-lighting arrangement for business use. Figure 3-14 shows continuous lighting alone.

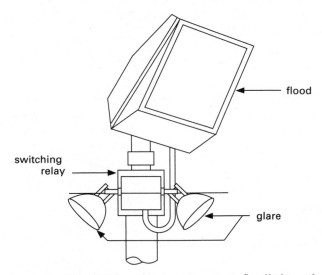

Figure 3-13. High-pressure sodium-type floodlight with glare projection.

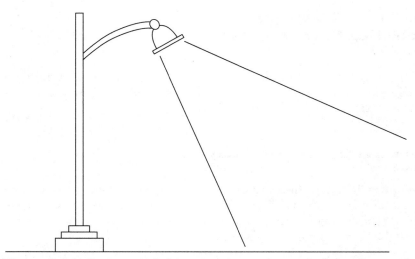

Figure 3-14. Continuous lighting arrangement on pole for parking and other areas. This arrangement can be fixed to a building.

Controlled Lighting. When it is necessary to limit the scope of your lighting (e.g., to prevent interfering with traffic or with other businesses), consider using controlled-lighting systems to protect your business at night. Controlled lighting is ideal for protecting the roof or exterior walls of your premises. When configured properly, it illuminates or silhouettes people moving about in a specific area. Figure 3-15 shows a roof application of controlled lighting.

Standby Lighting

Standby protective lighting is similar to continuous lighting, except that the luminaries do not stay lighted continuously. Instead, they operate by an automatic sensor that turns on the lights when movement is detected. The sensor is connected to an alarm system or other triggering mechanism.

Emergency Lighting

Emergency protective lighting may duplicate any or all of the above systems. Like the lighting used on construction or utility worksites, it serves you best during long-term power outages caused by storms, fires, or other problems. Emergency lighting relies on a backup power

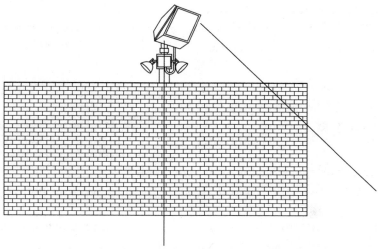

Figure 3-15. Controlled lighting arrangement. The example shows controlled lighting from the roof of a business with the glare lights serving to light the area.

source, such as portable generators or batteries. When the backup system turns on automatically, it should sound a clear alarm—one that is protected from being disarmed.

Protective Lighting Maintenance

You should make periodic inspections of all your business electrical circuits to repair or replace worn parts, tighten connections, and check insulation. Keep all luminaries clean and properly aimed. Lamps should be scheduled for group replacement at 80 percent of their rated life. Replaced elements can be used in less sensitive areas, or kept as emergency backups in case of breakage or other problems. However, be sure you label them with their expected life, and do not store them with new elements.

Remember that your emergency lines have actuating relays that remain open when your lighting system is operating from the primary source. These need to be cleaned regularly, since dust and lint will collect on their contact points and prevent them from operating in a closed position.

Security Technique 31: Plan a Protective Lighting System

When planning a protective lighting system, ask yourself these questions:

1. What are the costs and means (e.g., ladders and mechanical "buckets") of cleaning and replacing lamps and luminaries?

2. Are mercury and photoelectric controls advisable? These have the advantage of automatically turning on the system during weekends and according to levels of darkness, thereby eliminating the need for guessing or operating the system manually.

3. How will local weather conditions affect various types of lighting elements?

4. Are there fluctuating or erratic voltages in the primary power source?

5. Do fixtures need to be grounded, with a common ground placed on an entire line to provide a stable potential?

Security Technique 32: Select the Most Secure Lighting System

Before buying or installing protective lighting (or adding to or replacing existing lighting), consider the following tips:

1. Develop a full understanding of the manufacturer's descriptions, characteristics, and specifications of the various incandescent, arc, and gaseous discharge lamps available.

2. Determine the lighting pattern produced by each luminary, and compare it with your needs.

3. Design a layout of protective lighting on paper before buying and installing a system.

4. Determine the minimum lighting intensities needed to protect your business location during darkness.

Security Technique 33: Select Appropriate Fixtures

Once you understand the needs and characteristics of exterior lighting systems and have assessed the amounts of light your business requires, you need to find the appropriate fixtures. Remember that operating conditions—not a salesperson's opinions—determine your need for heavy-duty or general-purpose fixtures. In selecting the best possible protective lighting system for your business, decide what works for you and what fits your budget.

1. Will the fixtures be subject to rough use? (Consider heavy-duty items.)

2. Will the fixtures be exposed to severe atmospheric or corrosive conditions? (Consider heavy-duty items.)

3. Will your installation of protective lighting remain permanent? (Consider heavy-duty items.)

4. Will lightweight fixtures work and appear best in an area? (Consider general-purpose items.)

Here are some tips for deciding which elements to buy:

- Greater number of burning hours per year—iodine quartz
- Lesser number of burning hours per year—incandescent
- Color clarity—incandescent or iodine quartz
- Low initial floodlight cost—iodine quartz
- Location subject to power failures—incandescent or iodine quartz
- Inaccessibility because of positioning—iodine quartz
- Rapid relight capability not critical—vapor or fluorescent lamps

Quick Reference Guide to Chapter 3: Protective Exterior Lighting

Figure 3-16 (on page 64) presents a quick reference guide to the general principles of hardening your business security with protective exterior lighting.

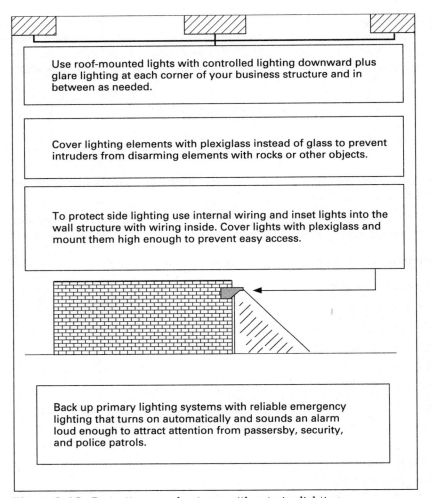

Use roof-mounted lights with controlled lighting downward plus glare lighting at each corner of your business structure and in between as needed.

Cover lighting elements with plexiglass instead of glass to prevent intruders from disarming elements with rocks or other objects.

To protect side lighting use internal wiring and inset lights into the wall structure with wiring inside. Cover lights with plexiglass and mount them high enough to prevent easy access.

Back up primary lighting systems with reliable emergency lighting that turns on automatically and sounds an alarm loud enough to attract attention from passersby, security, and police patrols.

Figure 3-16. Protecting your business with exterior lighting.

4

Choosing the Right High-Tech Security Devices for Your Business

The final layer of physical security for your business assets is to install an array of high-tech intrusion detection devices. Intrusion detection systems date back to about 390 B.C., when the Romans used squawking geese to alert them to a surprise attack by the Gauls. Today, trying to choose exactly what you need is a bit more complicated. The marketplace has become inundated with a myriad of devices—some inexpensive, others costly. When these devices are not properly understood, they can create a false sense of security.

What Is an Intrusion Detection System?

The basic electronic security systems for business application have three integral elements—including a sensor (integrated by data transmission and links) that detects the intrusion and signals a monitored annunciator console—backed up by a security response force. Intrusion detection systems are an inherent element of corporate and manufacturing security and play an important role in the total efforts of companies to protect their activities, information, equipment, and material assets. The systems detect intrusions through sound, vibration, motion, electrostatic means, light beams, and broken circuits. For an effective security effort, the system you choose must focus on detecting unauthorized persons at the vulnerable entry points (gates, doors, fences, and so on), at important areas (buildings and grounds), and at the specific objects (vaults, files, safes, and the like).

Remember that the intrusion detection system can tell you only that a protected area has been breached. It cannot do anything about the intrusion. Any system you choose depends on your ability to react positively to the alarm. Only adequate security or police officer response can keep the system functioning.

In this chapter, I will explain the basic elements of high-tech security devices, so that you will have an understanding of the products that goes beyond the manufacturers' claims. Remember that you need to decide what *you* want and what *you* need to protect your business assets. You also must exercise control over installation, because often the best intrusion detection devices fail when improperly installed. Don't leave these important decisions to others.

Benefits and Advantages of an IDS

Buying, installing, and using an intrusion detection system (IDS) as part of your total business security begins with a clear understanding of the benefits and advantages it offers.

Conservation of Financial Assets

An effective intrusion detection device often surpasses pure physical protection. Remember that eliminating the opportunity to commit a crime prevents the loss not only of merchandise, equipment, supplies, and other items but also of profits from disrupted business operations. With a solid nontechnological security system in place (i.e., locks, lighting, and structural hardening), as discussed in earlier chapters, plus preventive measures to be discussed in later chapters, intrusion detection devices are the capstones of asset protection and increased profits.

Substitutions for Other Hardening Techniques

Often lease restrictions, safety regulations, operational requirements, appearance factors, physical layouts, cost limits, and other factors prevent you from establishing adequate conventional security measures. The application of effective intrusion detection systems in those situations can solve protection problems that would otherwise be prohibitive.

Supplements for Other Countermeasures

Intrusion detection systems provide added control or layers of physical security for critical points of your business. Remember that high-tech security should always be used in relationship to basic security measures. An effective IDS supplies that final layer of protection, thwarting the most determined thief.

Factors Affecting Application of an IDS

Deciding to increase security with an intrusion detection system often becomes a problem of justifying cost. The following factors provide you with techniques of creating a cost-benefit formula to determine genuine need at your place of business. The formula will help you justify your decision and choose effective countermeasures.

Security Technique 34: Choose a System Appropriate to the Business

All businesses need security, but the exact amount and type depend largely on the nature of the operation. For example, a business producing massive steel construction beams (hard to pilfer) might need less asset protection than a business producing laptop computers (easy to steal and sell on the black market). In the steel business, it is unlikely that the manufactured product, construction beams, would be a target of criminals. Even if a thief tried to steal the beams, mustering the equipment to handle such items, doing so with concealment, and then selling the stolen property without attracting attention would be extremely difficult. However, the manufacturer still faces security problems with respect to other components of the business. Protection needs to focus on certain types of business information, small operating tools, supplies, and equipment in the steel manufacturing process.

Security Technique 35: Know the Vulnerability of Your Business Premises

Location plays an important role in security decisions. For example, if your business contains a lucrative product for employee pilferage or external theft, your first inclination might be to install heavy security based mainly on intrusion detection equipment. However, you need to consider the *likelihood* of intrusions and pilferage before making your decision. Businesses in a rural area traditionally need less security protection than similar businesses in or near any large metropolitan area (such as New York or Los Angeles) that experiences high levels of crime. Before installing elaborate high-tech security devices, carefully weigh the threat.

Security Technique 36: Assess Your Business Structure

As a rule, older buildings have more security problems than new buildings. Buildings constructed before air conditioning became standard have large windows for ventilation and other features that were not originally considered security threats. In this area of increasing crime, such structures face added threats and vulnerabilities. Newer buildings, too, are not immune. Many have structures designed for beauty, accessibility, and other features that present a variety of security problems. With those important factors in mind, choose your IDS according to the strengths and weaknesses of your business structure.

Security Technique 37: Evaluate Your Hours of Operation

Operating hours play a significant role not only in assessing the need and feasibility of intrusion detection devices but also in determining what type of system will serve your business best. For example, when your business operates 24 hours a day, 7 days a week, your primary security concern will clearly not be burglary. However, armed robbery and pilferage stay high on your list, especially if your business has several unstaffed storage warehouses. Examine all operational factors and plug them into your decision process.

Security Technique 38: Determine the Availability of Alternative Forms of Asset Protection

Depending on the type of business and the protection level you need, you might decide that intrusion detection devices do not fit into your total security plan. It is a mistake to install high-tech equipment just for the sake of having it on site. When you cannot clearly envision its need and feasibility, you probably do not need it. For example, if your business is in a shopping mall that has internal security officers during nonbusiness hours, plus regular police patrols on the outside, it will be less vulnerable to intrusion than a business that is in a freestanding building with only occasional police patrols passing by.

Security Technique 39: Assess the Initial and Recurring Cost of Devices

To determine the cost factor, consider first the adequacies of your basic security system and then figure the annual losses (including lost business) sustained by your operation (or others in the area) for the past 5 years. Next, determine the equipment cost, installation cost, and maintenance cost of adding intrusion detection systems.

Security Technique 40: Analyze the Design and Salvage Value

You need to consider the design feasibility of a specific intrusion detection system:

- How difficult is the system to operate?
- Is there a tendency for false alarms?
- Will the system need constant monitoring? (This is often an added expense.)
- Do you have enough time to manage and maintain the system?

   You also need to consider the salvage value of the system at later dates:

- Can the IDS be integrated effectively if you decide to upgrade your total security system later?
- Does the system you're considering have any value in the future if you decide to go out of business?
- Can you transfer the IDS to another business location or exchange it for another system better suited for your new business location, expansion, or operational changes?

Security Technique 41: Examine the Response Time by Security or Police Patrols

An intrusion detection system relies almost totally on the ability of security officers or police patrols to respond when a device alerts them to intruders on your business premises. When neither responds quickly, the system fails to protect your business assets. Thieves intending to steal property from a location that they know or suspect contains intrusion detection devices will often test the system. Typically, they trigger the system deliberately, then watch from a safe position to time a security or police response. The would-be intruders then determine the probability of completing their theft before security personnel can arrive—despite an operational and otherwise effective detection system. Criminals also might trigger the alarm system several times to create a track record of perceived "false alarms."

Security Technique 42: Estimate the Intruder Time Requirement

Evaluate the feasibility and cost of an intrusion detection system against the difficulty an intruder might experience entering your business premises. Without question, silent alarms at a business location offer the best choice—provided you can ensure a fast and adequate response by security or police patrol. However, an audible siren or bell also offers advantages, causing intruders to flee instead of steal. If you have sufficient security layers to confuse such intruders about how your system works, they will probably decide not to return later.

Basic Parts of an Intrusion Detection System

The four basic parts of an intrusion detection system are the sensor, control, annunciation, and response functions.

Sensor Function

The sensor function of an IDS detects or "senses" a condition that exists or changes, either authorized or unauthorized. The sensor relates directly to the human senses of touch, hearing, sight, smell, and taste. An example is the simplest and most common sensory device, the magnet contact on a door. This type of device activates with an alarm each time the door is operated. However, the sensor alone has no means of determining whether the person opening a protected door is authorized or unauthorized.

Control Function

The control function receives the information from a sensor, evaluates the information, and transmits it to the annunciation function. The control function compares loosely with the brain, nervous system, and circulatory system of the human body. The nervous system collects and evaluates information from the various parts of the brain (sensors) and transmits signals to the muscles for appropriate action. The circulatory system then supplies the power source (i.e., nutrients and oxygen from the blood) to maintain the system's ability to function.

Annunciation Function

The annunciation function alerts a human monitor to initiate an investigation of the sensor environment (such as notifying the police or a security agency). The annunciator might be a bell, buzzer, or flashing light. This function is similar to the squawking of geese, the barking of dogs, or a person calling for help.

Security or Police Patrol Response

Thus far, the intrusion detection system has identified a change of conditions, evaluated the information, and articulated a call for help. The fourth function brings in the critical factor of physical response. Without the human element, even the most sophisticated IDS is ineffective.

Problems with Intrusion Detection Systems

Although intrusion detection systems have proven effective in deterring and apprehending intruders, they also have inherent problems.

The False Alarm Factor

A key problem with IDS applications is the high percentage of false alarms, which can be traced to three key factors:

- User error or negligence, including lack of preventive maintenance.
- Poor installation, or placing the system in a location that is incompatible with its purpose.
- Faulty or poor-quality equipment

Statistics show that more than half of all IDS false alarms are caused by user error or negligence. Users often do not understand how to operate their systems. Commonly, false alarms occur when employees do not lock doors or windows, or enter a secured area without realizing that an alarm system is engaged. Alarm systems may also be activated accidentally by workers and cleaning crews, and by outside noise and vibrations.

Poor installation or servicing also poses problems. Equipment that is installed in an inappropriate environment, or that is improperly positioned, set, or wired, produces false alarms. If equipment is not adequately maintained, the chance of false alarms increases as well. Too often, installers and service people lack the necessary skills and knowledge to operate sophisticated IDS equipment.

Faulty or poor-quality equipment is the third culprit. When intrusion detection equipment is electrically or mechanically defective (e.g., the equipment breaks or shorts out the circuit), the alarm may activate. Cheap or substandard equipment is especially vulnerable to breakdown and can be easily set off by a variety of extraneous conditions.

There are also a number of undetermined reasons for false alarms. Studies conducted over the years show that on average 25 percent of IDS false alarms fall into this "unknown" category. It is possible that they are the result of undetected user error. Another possibility is that a potential criminal was warned by the alarm and fled, leaving no visible evidence of intrusion or attempted entry. Acts of nature (storms and earthquakes) can also trigger an alarm system.

Compounding Factors

Whatever their cause, frequent false alarms lead to other problems. Automatic telephone dialer alarm systems are particularly vulnerable. Many storm-related false alarms happening simultaneously may tie up telephone lines and seriously hamper the ability of security or police to respond to genuine situations. Malfunction of the IDS can lock out communications systems for considerable periods of time. Although telephone dialers offer effective, low-cost protection, these problems have created negative reaction to their use.

Police attitudes about intrusion alarm systems are generally less than favorable. Faced with repeated false alarms, police officers tend to give alarms a low priority of response. The resulting delay reduces the likelihood of apprehending intruders and limits the value of intrusion detection systems. Further, police officers, lulled by the high incidence of false alarms, may not conduct thorough on-the-scene investigations or be alert to the risks of valid alarms.

Solutions to IDS Problems

As noted earlier, intrusion detection systems can provide you with the valuable advantage of complete security. They offer added protection at a cost lower than that of salaried security officers, and often reduce the high insurance premiums associated with high rates of crime.

Remember that reduced losses equal increased business profits. With all the benefits and advantages of an IDS, the solution becomes a cooperative effort of all parties to overcome existing problems. These important "others" include IDS manufacturers, dealers, sales and service people, business employees, and local law enforcement officers who may respond to the alarm. The interdependence and interaction of these groups are critical to the system's reliability and effectiveness.

Security Technique 43: Work Closely with Manufacturers

Performance standards for separate IDS components and installed systems need constant improvement, and you can help with recommendations and documented problems. When you document problems and share the experience with the manufacturer, weaknesses can be overcome.

You can also help by working closely with manufacturers to test their improvement in real circumstances. In the attempt to develop more reliable human-discriminating sensors, you can play a key role with your suggestions and save your business significant sums of money in the process.

Assuming you have dependable equipment installed in a compatible environment by well-trained and well-qualified personnel, there is one more factor you need to consider: user error or negligence. Here, too, the remedy requires effective cooperation with the manufacturer. You need to train yourself and educate your employees about systems installed in specific areas. Often, IDS users assume important responsibility for proper operation and total effectiveness, but are unaware of essential operating procedures or the effect of false alarms.

Security Technique 44: Develop Backup Power by Coordinating Business Operations

Ensure that you arrange for effective backup power sources for your intrusion detection systems. An important problem with power sources is severe weather conditions that interrupt power from a few seconds to several hours. Heavy drains on power, breakdown of power stations, fires, motor vehicle accidents that involve striking a pole and downing wires, and other factors can effectively shut down your intrusion detection systems. Consider using two backup power sources to ensure adequate protection. Backup sources include batteries and portable or stationary generators. Batteries provide immediate continuance of operation for several hours. However, batteries are not enough. Emergency generators are useful but costly, especially when you need significant amounts of power over long periods of time.

It is best to develop backup power through cooperating with other business operations. For example, without power to continue functioning, downed computers will create major business problems. Your security lighting, intrusion detection systems, and other heavy power needs, are necessary primarily during the hours of darkness. In the daytime, your business can function on workaround measures not dependent on electrical power. With this concept in view, your business manager probably will agree with you that a costly backup system supplying electricity for essential business operations during the day and enabling effective protection of business assets at night is a good investment.

Overextension of IDS Equipment

The problem of overextension develops when a system works effectively under ideal conditions, but does not protect a business when conditions are not ideal. Typically, it results from attempts to economize or from allowing a manufacturer or supplier to tell you what you want or need. The solution is to carefully evaluate the system offered. You should reduce the manufacturer's claimed protection benefit by about 30 percent!

Security Technique 45: Use Sensor Devices Effectively

Devices for the sensor function of an alarm system provide a broad range of choices, from simple magnetic switches to sensitive ultrasonic Doppler, light beams, and sound systems. Here are the major systems now available.

Alarm Glass. Small wires molded into glass form an electrical circuit that triggers an alarm if the glass is cut or broken.

Alarm Screens.   Alarm screens are placed in front of windows. The screen mesh contains thin wires that establish an electrical circuit. The alarm is set off if the screen is cut, thereby interrupting the circuit.

Capacitance Proximity Detectors.   Proximity sensors respond to a change in the electrical capacitance of a protected metallic item or area and are activated by the approach of an intruder. They are used on metallic screens, doors, safes, cabinets, and similar items.

Duress Alarms.   Duress devices activate manually and alert security or police officers to a situation requiring their response.

Metallic Foil Sensors.   Foil sensors are used on glass or other breakable material, windows, walls, or doors to detect the breaking of the glass or other material. When the material is smashed, the thin foil breaks, cutting an electrical circuit and activating an alarm.

Photoelectric Controls.   Photoelectric controls sense a change in available light either from a light source or from the general area.

Photoelectric Sensors.   Photoelectric sensors project a light beam or invisible infrared beam, triggering an alarm when an object or person interrupts the beam.

Pressure Mat Sensors.   Pressure sensors are placed under interior or exterior carpets or under grass and other materials. They trigger an alarm when a given level of pressure is applied—for example, when an intruder walks across the pressure mat.

Pull-Trip Switches.   Mechanical pull-trip sensors respond to the pulling of a wire connected to a protected object. For example, a thin wire or maze of wires placed across an area can activate the switch when an intruder walks into it; or wires attached to a door or other object will activate a switch when the object is opened. Pull-trip switches work best for doorways and other entrance points.

Stress Detectors.   Stress-sensitive instruments connect to a floor beam, stairway, or other structural component. They activate an alarm by detecting a slight stress placed on the component.

Switch Sensors.   Switch sensors detect the opening of doors and windows. There are three main types.

- *Magnetic switches.* The contacts of the switch open and close in response to the movement of a magnet mounted on it.
- *Mechanical switches.* The contacts open and close by spring action in response to the opening and closing of a door or window.
- *Mercury switch sensors.* The contacts open and close when the switch tilts. These highly sensitive devices protect doors, windows, transoms, and skylights.

Vibration Sensors. Vibration sensors detect vibrations in walls, floors, or other structural elements of a building and are activated when force is used to penetrate the area protected.

Infrared Motion Sensors. Infrared devices sense a source of heat that moves into their field. Even the body heat of an intruder is enough to trigger an alarm.

Microwave Motion Sensors. Microwave sensors detect motion in the protected area and trigger an alarm.

Sound-Monitoring Sensors. Sound sensors trigger alarms from sharp noises, and serve as listening devices for security officers at a control station.

Ultrasonic Motion Sensors. Ultrasonic sensors generate a high-frequency sound above the range of human hearing. Motion in the protected area triggers an alarm.

Laser Sensors. Laser systems are similar to photoelectric sensors, except that laser beams replace light beams.

Wafer Switch Sensors. Wafer switches are small, negative-pressure switches. Instead of signaling an alarm under pressure, they operate when the pressure is released. They are used to protect objects that an intruder might pick up as part of a theft and offer some protection against illegal entry to collect competitive intelligence.

Doppler Sensors. The Doppler system works from a continuous emission of microwave beams (much like Doppler radar systems installed in police cars to detect speed). The alarm triggers when a moving object (i.e., a person) enters the area and compresses the beams.

Security Technique 46: Evaluate the Compatibility of Sensors

Carefully study the environment you intend to protect before purchasing sensors. Overlooking certain environmental factors at the time of installation can make the system useless. For example, installing vibration detectors in a bank vault above a subway would be futile. Factors such as the state of repair of a building and the type of heating (radiators, air currents, and so on) can affect the successful operation of sensory equipment. You must ensure that the sensor devices are operationally compatible with the area where they are located and that they provide the intended results.

Security Technique 47: Set Up an Economical and Effective Intrusion Detection System

A simple and inexpensive way to protect points of entry into your buildings or enclosures is to use electrically sensitized strips of metallic foil or wire. Any action that breaks the foil or wire also breaks an electrical circuit and activates an alarm. Metallic foil works effectively on glass surfaces and protective screening. Doors and windows equipped with magnetic contact switches sound an alarm when the door or window is opened. Metallic wires running unseen through wooden dowels or between panels or walls, doors, and ceilings can also provide adequate protection.

Advantages.   Electrical circuit systems consistently provide the most trouble-free service. They cause few, if any, nuisance alarms, and provide adequate protection in low-risk environments. Often they are the most effective and economical IDS devices.

Disadvantage.   When a protected area has many openings, installation of electrical circuits may be complicated. Electrical circuit systems are especially susceptible to compromise in unprotected soft walls or ceilings that can be penetrated in ways that bypass the system. Defeating the system by bridging the electrical circuits is not foolproof. The system uses magnetic contact switches recoverable for use elsewhere. It cannot detect "stay behind" intruders, unless they try to leave the area or building.

Security Technique 48: Add Redundant Systems

Electrical circuits are most effective as primary systems in buildings or areas that do not have high-value items, or as support devices for sensitive areas protected by other systems. For example, a combination of visible and hidden electrical circuit systems serves as both a prevention feature and a detection feature.

Unseen systems have the same effect as props used by a magician. The entertainer makes certain elements clearly visible and draws the viewer's attention to them, while using unseen elements to create an illusion. A clearly visible system, in and of itself, will deter many burglars, especially amateurs. Redundancy becomes effective when a skilled intruder carefully bypasses the obvious system and is detected by concealed circuitry. Figures 4-1 through 4-3 show applications of electrical circuit systems.

Security Technique 49: Use Invisible Light Beams

The photoelectric (electric eye) beam relies on a light-sensitive cell and a projected light source to detect intrusions. This type of protection operates as follows:

1. A light beam is transmitted at a frequency of several thousand vibrations per second. An infrared filter over the light source makes the beam invisible to intruders.

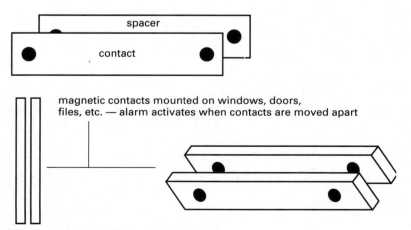

Figure 4-1. Magnetic contact switches.

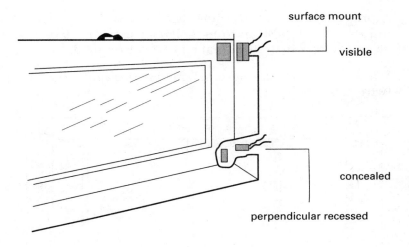

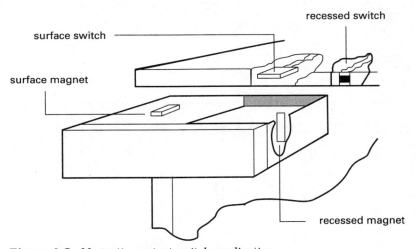

Figure 4-2. Magnetic contact switch application.

2. A light beam with a different frequency (such as a flashlight) cannot be substituted for the photoelectric beam. The beam projects from a hidden source and crisscrosses the protected area via hidden mirrors until it contacts with a light-sensitive cell.

3. Wires connect the light-sensitive device to a control station. An intruder crossing the light beam breaks contact with the photoelectric cell, and that activates an alarm.

4. A projected beam of invisible light is effective for about 500 ft indoors and 1000 ft outdoors. The effectiveness of the beam decreases from 10 to 30 percent for each mirror used.

Figures 4-4 and 4-5 show exterior and interior applications of a photoelectric intrusion system.

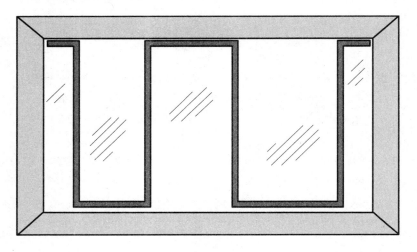

(a)

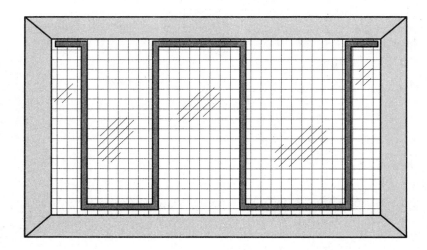

(b)

Figure 4-3. (*a*) Metallic foil on glass window (*b*) coupled with protective mesh screen.

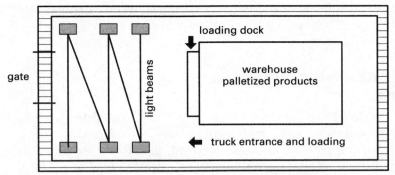

Figure 4-4. Double-chain-link fence with protective lighting.

Advantages.   A photoelectric system yields effective, reliable notice of intrusion. It works well in open portals or driveways where there are no obstructions and effectively detects "stay behind" intruders.

The photoelectric system increases in value because it can be recovered and used elsewhere. It can also actuate other security systems, including cameras and lighting. Finally, the system may detect fires when smoke interrupts the beam.

Disadvantages.   A photoelectric system is effective only in locations where intruders cannot bypass the beam by crawling under or climbing over it. The system needs some type of permanent installation. Also, factors such as dense fog, smoke, dust, and rain can interrupt the light beam and trigger the alarm. The system calls for frequent inspection of light-producing components for deterioration. The ground

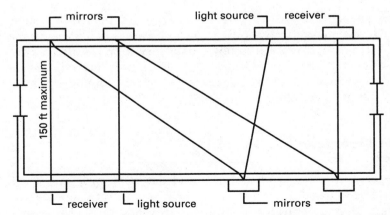

Figure 4-5. Photoelectric beam interior application for hallway, entrance, or room.

beneath the light beam must be free of tall grass, weeds, drifting snow, and sand.

Sound-Activated Detection Devices

Sound-activated intrusion detection systems effectively safeguard vaults, warehouses, and similar enclosures. Supersensitive microphone speakers installed on walls, ceilings, and floors detect sounds made by forced entry. The system can be adjusted by regulating the levels of sound needed to trigger the alarm.

Security Technique 50: Create a Sound Detection System

You should consider using sound-activated devices as a clandestine listening method to support other intrusion detections. For example, when a sound triggers the alarm in an area where other systems should also respond but do not, security officers in a control center can listen to determine whether the alarm is false or real. An intruder might bypass these other systems without knowing that a sound detection program is also present. Thus, it is an excellent "layer" or redundant system to install.

Advantages.   A sound detection system is economical and easily installed. You can recover the components from one location and install the system in another building or area as needed.

Disadvantages.   The greatest limitation of a sound detection system is noise. The system is effective only in enclosed areas that have a minimum of extraneous sound. It is not satisfactory in areas affected by high noise levels—for example, areas close to aircraft and railroad or truck traffic—or in outdoor areas, where extraneous sounds will activate false and nuisance alarms.

Security Technique 51: Detect Vibration

Vibration systems work well with a sound detection program. They are effective in closed areas and activate when an intruder creates vibrations by forcing entry or moving about. Vibration-sensitive sen-

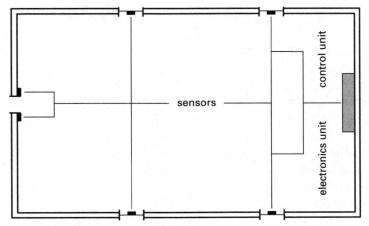

Figure 4-6. Vibration sensor application.

sors are attached to walls, ceilings, and floors of the protected area. Any vibration triggers the alarm system. Figure 4-6 shows a vibration sensor application in a business building.

Advantages.   Vibration systems are economical and easy to install. Their components are recoverable and easily moved to other locations. With their highly flexible applications, vibration devices are effective as independent systems or as support systems for other layers of security.

Disadvantages.   Vibration systems can be used only in areas where a minimum of vibration exists. They are not satisfactory or reliable outdoors or in areas where high vibrations are encountered, especially in proximity to heavy construction, railroads, or automobile and truck traffic.

Security Technique 52: Detect Motion Ultrasonically

Intrusion detection systems using ultrasonic motion sensors can be very effective in protecting interior areas. Such systems flood the area with acoustic energy and detect the Doppler shift in transmitted and received frequencies when motion occurs.

Ultrasonic systems have a transceiver (a single unit containing a transmitter and receiver) or a separate transmitter and receiver, plus an electronic unit (amplifier) and a control unit. The system operates as follows:

1. The transmitter generates a pattern of acoustic energy that fills the enclosed area.
2. The receiver, connected to the electronic unit, picks up the standing sound patterns.
3. If the sounds are of the same frequency as the waves emitted by the transmitter, the system will not activate the alarm.
4. Any motion within the protected area sends back a reflected wave differing in frequency from the original transmission. The change in frequency is detected and amplified, and the alarm signal is activated.
5. Multiple transmitters and receivers can be operated from the same control unit for more effective coverage of large or broken areas. This system works only indoors.

Figure 4-7 shows the operation of an ultrasonic device.

Advantages.   An ultrasonic system provides effective detection of intruders hidden on the premises. The system has high salvage value—components can be recovered and used in other locations. Since the protective ultrasonic field is not visible, it is difficult to detect the system's presence or to compromise its operation.

Disadvantages.   An ultrasonic system is highly sensitive. Adjustments may be required to overcome possible disturbances in the enclosed area (e.g., from telephones, machines, and clocks). The system can be set off by any loud external sound.

An ultrasonic motion sensor activates an alarm even when there is only a flow of air from heating units, air conditioners, cracks in the protected area, or any other possible source of air turbulence. This reduces the effectiveness of the sensor and may cause nuisance alarms.

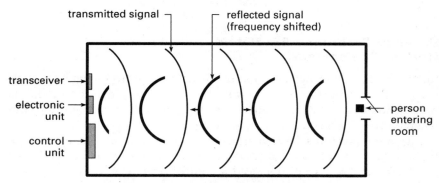

Figure 4-7. How an ultrasonic motion sensor works.

Security Technique 53: Establish Microwave Systems to Detect Motion

Microwave systems closely parallel ultrasonic systems in detecting motion. A pattern of radio waves transmits back to an antenna. If all objects within the range of the radio waves are stationary, the reflected waves return at the same frequency. However, if the waves strike a moving object, they return at a different frequency. The detected difference in the transmitted and received frequency triggers an alarm signal.

Advantages.   Microwave systems offer good coverage if antennas are properly placed. They are not affected by air currents, noise, or sounds. Components can be recovered from one location and installed in another.

Disadvantages.   Fluorescent light bulbs will activate the microwave sensor. Also, coverage cannot be confined to a desired security area. It penetrates thin wooden partitions and windows and may be accidentally activated by people or vehicles outside the protected area.

Security Technique 54: Use Electrostatic Field Systems

Electrostatic intrusion detection systems perform effectively when installed on a safe, wall, or opening to establish an electrostatic field around the protected object. The field is tuned to create a balance between the electric capacitance (ability to store an electric charge) and the electric inductance. The body capacitance of an intruder entering the field unbalances the electrostatic energy, activating an alarm system.

Figure 4-8 shows applications of electrostatic systems.

Advantages.   An electrostatic system is extremely flexible. It can be used to protect safes, file cabinets, windows, doors, partitions, and any unguarded metallic object within maximum tuning range. The invisible protective field makes it difficult for an intruder to determine when the system will activate. Electrostatic equipment is compact, simple to install and operate, and easily dismantled and reinstalled at a different location.

Disadvantages.   Electrostatic techniques can be used only with ungrounded equipment, and protected areas need careful housekeeping. Accidental alarms can occur if the protected area or object is approached carelessly, especially by cleaning crews or others working after business hours.

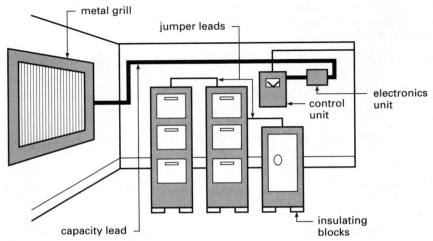

Figure 4-8. Electrostatic sensor system.

Security Technique 55: Use Penetration Sensors

Modern penetration alarm sensors serve well as an added security layer at any business location. They are versatile and can be used on windows, interior or exterior doors, ceilings, walls, and other potential entry areas. Here are several examples of their application.

Exterior and Interior Doors. To guard against unauthorized entry, equip the door with balanced magnetic switches. Cover the surface of an interior door or wall with a grid-wire sensor or any type of system that uses the principle of breaking an electrical circuit to warn of intrusion.

Solid Walls, Floors, and Ceilings. To monitor attempts at penetration of solid walls, floors, and ceilings, cover the interior surface with a grid-wire sensor, and equip the protected room with a passive ultrasonic sensor. Sound detection systems and vibration detection systems work effectively to detect penetration through such areas.

Open Walls and Ceilings. Wire-cage walls and ceilings present distinct problems. To protect this type of construction, certain modifications are necessary. The surface wall and ceiling outside the cage should be enclosed with building material to permit the use of passive ultrasonic or grid-wire sensors.

Windows. Whenever possible, eliminate windows and other openings. These supply lucrative entry points for intruders. When windows are necessary, consider using interior metal shutters that can be closed and locked, to permit the use of passive ultrasonic sensors. If the room does not allow for installation of a passive ultrasonic sensor, a vibration sensor or capacitance proximity sensor will suffice. Also, consider a system based on the principle of breaking an electrical circuit.

Ventilation Openings.　Any openings in the ceiling, walls, or doors that allow the free passage of air need to be blocked with steel bars, mesh, or louvered barriers. For maximum protection, you should eliminate ventilators. The second best approach is to install locked metal shutters to permit use of a passive ultrasonic sensor or vibration sensor. When ventilators are open continually, place a metal grill over the ventilator opening and attach a capacitance proximity sensor to protect it.

Construction Openings.　Unsecured openings from unfinished construction should be covered with a grid-wire sensor installed on plywood. If the opening must stay open, use a capacitance proximity sensor within it.

Air Conditioners.　To monitor for intrusion through an air-conditioner aperture, attach a capacitance proximity sensor to a metal grill extending into the room in front of the unit.

Figures 4-9 and 4-10 provide examples of balanced magnetic switches placed on exterior doors, with grid-wire sensors used on interior surfaces. Figure 4-11 illustrates passive ultrasonic sensor applications.

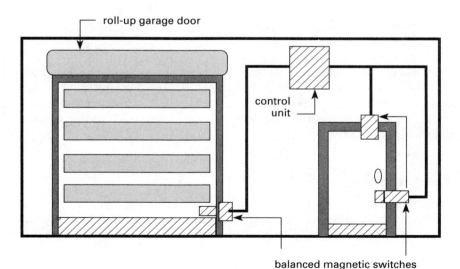

Figure 4-9. Balanced magnetic switches placed within exterior doors.

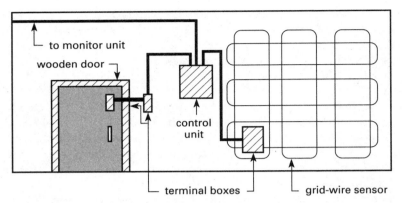

Figure 4-10. Grid-wire sensors used on interior surfaces.

Other Detection Devices

Duress sensors, point sensors, data transmission systems, and alarm detection systems complete the battery of devices available to secure your business. Each offers advantages in certain situations.

Duress Sensors

Fixed.   The fixed duress sensor alerts security officers that there is an employee requesting immediate help. It activates from a foot- or hand-operated switch placed in a position that is easily accessible to employees working in the protected area. Figure 4-12 presents an example.

Portable.   An alternate to the fixed duress sensor, the portable duress sensor is an ultra-high-frequency (UHF) transmitter that can send an alarm to a receiver at the control unit. The restricted effective range depends on whether the sensor is used inside a building or outside in an open area.

Point Sensors

Monitor Units.   To detect movement near to, or discover contact with, any part of a storage cabinet or safe, you need to use a capacitance proximity sensor with control and monitor units, plus an audible alarm.

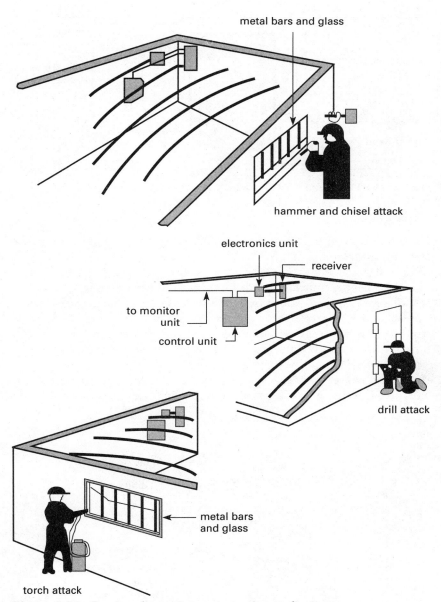

Figure 4-11. Passive ultrasonic sensor system applications.

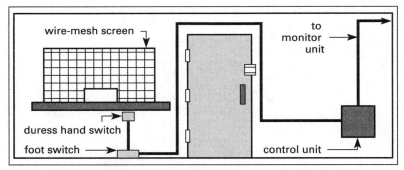

Figure 4-12. Placement of a fixed duress sensor for a financial disbursement center.

Place one control unit in each secure area to receive signals from the sensors and transmit signals elsewhere in the system (see Fig. 4-13). Each control unit must report to a separate monitor unit (see Fig. 4-14), which contains one signal indicator module and one power status module.

Install a local audible alarm outside the protected area to scare intruders away and to alert security officers in the area (see Fig. 4-15). This alarm has limited value, of course, if no response teams are assigned to the area. A local audible alarm works best with a remote monitor unit.

Telephone Dialer. A telephone dialer might serve your needs if it is not possible to install a monitor unit. This device telephones an

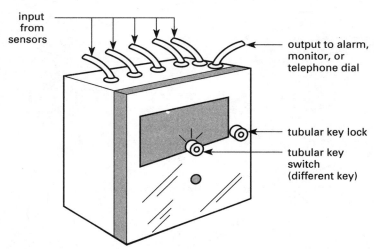

Figure 4-13. Intrusion detection system control unit.

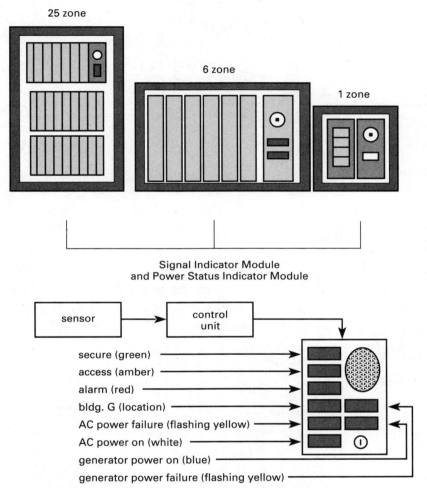

Figure 4-14. Monitor unit for intrusion detection systems.

alarm to several preselected phones. Use telephone dialers only for low-security applications. Their lines can tie up when the system continues to call the numbers for alarm notification. Telephone dialers are subject to tampering and interruption and do not activate when the lines are out of order, cut, or grounded. Figure 4-16 illustrates a telephone dialer system.

Data Transmission Systems.   Data transmission systems work best when segments of a signal transmission line are accessible to tampering, or when the signal is sent over commercial lines. One data

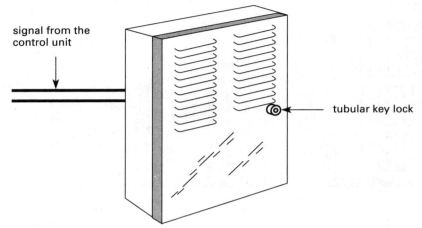

Figure 4-15. Audible alarm.

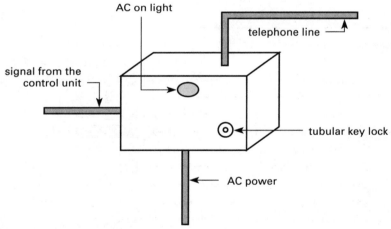

Figure 4-16. Telephone dialer system.

transmission system should be installed in each security zone covered by a control unit and must be connected to the control unit. Figure 4-17 illustrates a data transmission system.

Alarm Detection Systems

Alarm detection and communications systems become closely allied in any comprehensive protection program. Telephone and radio commu-

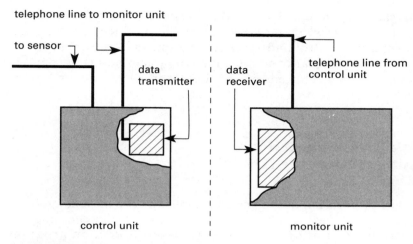

telephone line to monitor unit

to sensor

data transmitter

data receiver

telephone line from control unit

control unit

monitor unit

Figure 4-17. Data transmission system.

nications are so common in business that their adaptation to a protective system poses few new problems. An alarm detection system is simply a manual or automatic means of communicating a warning of potential or present danger. There are several types of alarm systems.

Auxiliary System. A business-owned auxiliary system is a direct extension of the area police and fire systems. It is the least effective alarm system because of the dual responsibilities for maintenance.

Local System. In a local alarm system, the protective circuits or devices actuate a visual or audible signal in the immediate vicinity of the object or protection. Response is made by the security force that is within sight or hearing range. The alarm device must be installed outside the building and must be protected against weather or willful tampering. The component connects to the control element with a tamperproof cable and is kept visible or audible for at least 400 ft. This type of protection can work with a proprietary system (see below).

Central Station System. A commercial agency may contract to provide protective services to its clients by use of a central station system. The agency designs, installs, maintains, and operates underwriter-approved systems to safeguard against fire, theft and intrusion, and monitors the manufacturing processes. Alarms transmit signals to a central station outside the corporate or manufacturing complex (such as a local police or fire department), where appropriate action can be taken.

Proprietary System.   A proprietary system is similar to a central station system except that it is owned by the proprietor on the business property. Control and receiving equipment are located in the business security department. This type of system normally connects with the local law enforcement and fire departments through a commercial central station.

Quick Reference Guide to Chapter 4: Six Key IDS Factors

Each type of intrusion detection system must meet a specific problem within your business. Factors you need to consider in choosing the best components or systems for your asset protection program include, but are not limited to, the following six points.

- *Response time capability of a security or police patrol.*   Response time is the critical factor in evaluating the value of an intrusion detection system. Without timely human response, your system will not create any advantage.

- *Value of protected business facility and assets.*   Evaluate the cost of the system opposed to the value of assets and other factors you protect.

- *Business environment, including outside and inside factors.*   Evaluate the outside and inside environment of your business. Include factors that can trigger alarms from vibration, wind, or other elements.

- *Radio and electrical interference.*   Evaluate the probability of radio or electrical waves triggering the intrusion detection system.

- *Operational hours of the business.*   Consider how many hours your business is available to potential intruders. For instance, if your business operates around the clock and closes only 2 days a year, a burglar intrusion system would not have a cost-effective benefit.

- *Specific business facility, area, or site protected.*   Each business facility, area, or site will have different requirements. Choose the system that meets your specific standards instead of a system that claims to serve any purpose.

5

Eliminating the Threat of Armed Robbery

The crime of armed robbery has caused businesses not only to lose assets but even to close down. Robbery is a traumatic personal experience for owners, managers, employees, and customers. Many have been injured or killed by assailants. The rate of armed robbery has increased steadily and significantly over the past 20 years. The repercussions of armed robbery can work themselves into the fabric of your business, and the accompanying publicity can lead to a marked slowdown in customer activity. People are afraid even when no robbery has happened, because everyone fears that it might happen one day.

The good news, however, is that you do not have to become a robbery victim. Use the key technique noted throughout this book—eliminate the opportunity—and you will prevent the crime. This rule must always guide your thinking as you search for ways to end crimes such as armed robbery.

In this chapter, I will first lead you through a definition of armed robbery. The elements of robbery must be understood before you can devise effective countermeasures. Then I will supply you with proven techniques of prevention that eliminate the opportunity and make the risks too high for the offender. Finally, I will show you how to implement your solutions without interrupting business operations.

It's important to remember that people who commit armed robbery do not plan to become statistics in police records. Instead, they intend to escape to enjoy the fruits of their crime. Knowing that, you need to understand the threat from all categories of armed robbers.

Armed Robbery:
A Definition

Many businesspeople remain confused about what legally constitutes a robbery. They often refer to burglary or another crime of theft as "being robbed." Robbery is not just a crime against property but also a crime against person (or business), with a strong potential for physical violence.

The use or threat of force is a critical element in defining robbery, and the force must be sufficient to make a reasonable person fearful. In this sense, the line between theft and robbery is sometimes thin. For example, if a person enters your business, grabs something of value, and runs away so quickly that you cannot offer any resistance, the crime will legally be considered theft or larceny. However, if you struggle with the offender, the act will more likely be classified as robbery, especially if the person is of superior strength or threatens you with a knife, gun, or other object that could obviously injure or kill you. The crime of robbery will often have legal classification according to the degree of force used or threatened by an offender.

The FBI defines robbery as "the taking or attempting to take anything of value from the care, custody, or control of a person or persons by force or threat of force or violence and by putting the victim in fear." *A robbery is a crime of violence because it involves the use of force to obtain money or goods.*

Security Technique 56:
Understand the Armed Robbery
Offender_____

Sometimes criminals decide on robbery only a few minutes before they act. These offenders piece together the elements of a crime by improvisation. Government and law enforcement studies of armed robbery over the years suggest that more than 40 percent of juvenile and 25 percent of adult robbers did not intend to commit the robbery before it occurred.

However, certain robbers do "case" their business targets. They may look at the architecture of a business and its location. For example, a streetcorner business near a superhighway affords a fast getaway. These robbers also assess the risk, taking into account the number of witnesses likely to be at the scene. In trying to neutralize obstacles, robbery offenders often look at the crime from the victim's perspective. They calculate how the victim will react to the crime and how important the money they steal is to that victim.

Often, planning a robbery includes an assessment of the best time to commit the offense. The offender chooses a time that will maximize the score and minimize the obstacles. Businesses that pay employees in cash or that draw customers by offering to cash their payroll checks become targets because they have large sums of money on hand. Some armed robbers prefer to commit their offenses on rainy days, because they can wear a coat to conceal weapons and will not seem suspicious on the street if they run from the crime scene through the rain.

Robbers also may choose to act when the streets are empty of people and cars, making their escape easier. They may watch the flow of people into an establishment—counting customers in a store or a bank on the same day and at the same time—for weeks or months ahead of a planned robbery. They may try to figure out when messengers carry money to night depositories or determine when store managers leave the premises.

Security Technique 57: Learn the Four Primary Categories of Robbery Offenders

Different categories of robbers pose different types of threats. Learning to recognize the categories will help you assess your vulnerability and find effective countermeasures to eliminate robbery opportunities.

Professionals

Professional robbers include those who manifest a long-term commitment to crime as a source of livelihood, those who plan and organize their crimes before committing them, and those who seek money to support a particular lifestyle. Some professionals are exclusively robbers, whereas others also engage in other types of crimes. Professionals commit themselves to armed robbery because it supplies a direct, fast, and ready way of getting money. Most hold no steady legitimate job and plan enough robberies each year to support the type of lifestyle they seek. Normally, professionals operate in groups, with each participant assigned a specific role in the crime. They prey on local businesses, especially those with large amounts of cash or merchandise that is easily converted to cash.

Opportunists

Unlike professional robbers, opportunists commit a variety of property crimes besides robbery. Robbery is an infrequent activity for them. In contrast to professional robbers, they rarely score in a big way, and the small amounts they do obtain generally support only spur-of-the-moment needs that help them maintain their lifestyle or peer group image. Their robberies are unplanned, haphazard affairs, and their decision to rob is often an impulse, influenced by momentary pressures and the availability of victims. The opportunist will execute impulsive attack robberies on small businesses.

Drug Users

Drug-influenced robbers include those who are addicted to drugs and steal to support their habit, as well as those who simply use drugs, most notably pills such as amphetamines and psychedelics. In the latter case, the robberies may or may not be motivated by the need for cash to buy more drugs. Drug users may see robbery as a steady way to finance the purchase of illicit drugs for resale. They may take cash from the business owner, employees, and customers and gain more cash by stealing prescription drugs for resale. Drug addicts, by contrast, usually have a low level of commitment to armed robbery. They consider it a last resort because of the high risk and danger involved. Most commit their crimes impromptu, out of desperation, instead of carefully planning them.

The extent to which robbery is solely drug-related is hard to establish. A variety of studies by government agencies suggest that other property offenses (such as burglary, shoplifting, and theft) have greater appeal for those involved with drugs.

Alcoholics

Many people steal for reasons related to excessive consumption of alcohol. They have no real commitment to robbery as a way of life and plan their crimes randomly, with little thought to victims, circumstances, or escape. When a crime displays this pattern, the police typically begin their search at nearby bars or hangouts and are able to find alcoholic armed robbery offenders quickly and easily.

The Anatomy of Armed Robbery

While professional armed robbery offenders are less numerous than their amateur counterparts, they have received more attention and have been investigated more systematically. Numerous studies have asked professional robbers how they work. We know more about their activities, work attitudes, and organization than we do about those of amateurs. Of course, many amateurs mimic professionals as they climb up the career-criminal ladder.

Since professionals make robbery their career, they cannot afford to fail or get caught. They emphasize attentive planning, anticipation of difficulties, assessment of risks, and teamwork. To get some idea of how career robbers and amateurs who mimic them pursue their work, let's look at five typical phases in the robbery cycle:

- Going into partnership
- Planning and making arrangements
- Reconnaissance for the robbery
- Conducting the robbery
- The getaway

Going into Partnership

First-time armed robbery offenders, and even those who have taken part in robberies before, must often get a team together before embarking on a robbery or series of robberies. What motivates first timers to get into this particular line of work? It could be on the invitation of an acquaintance who has experience in armed robbery and needs a partner. It could be the conscious decision of experienced thieves to enter a new "line" of work. Sometimes the decision to get involved in armed robbery comes after a careful assessment of what best serves a person's needs.

Planning and Making Arrangements

Once a partnership is formed, the partners spend most of their time and resources planning for the crime. Many arrangements have to be made. Those who have been in the profession for some time will make sure they have "contingency people" in the wings—such as discreet doctors to aid with injuries, lawyers to handle trouble with police, and bondspeople to pay bail money. When the robbery involves securities, large amounts of cash that authorities might trace, or jewelry or other items that need to be converted to cash secretly, the group will find the right "fence" to handle the transactions, usually in a distant city.

The planning phase of the robbery may differ from one partnership to the next, but certain activities will remain the same. Among the basics that the team must establish are:

- The business to be robbed (bank, supermarket, liquor store, and so on)
- Role assignments for the robbery: driver of the getaway car, lookout, primary armed robbery offender (normally the senior criminal), and assistants to the primary offender.
- Stolen cars, false license plates, and routes to and from the robbery
- Dates and times for "casing" or observing the targeted business and making dry rehearsals.
- Place and time for splitting up the money or merchandise.

Reconnaissance for the Robbery

Every member of the robbery team participates in reconnaissance, or "casing" the targeted business. The team drives around the area, noting parking opportunities, entrances and exits, police patrols, and the movement of people into, out of, and around the business. Sometimes special attention is given to the architecture of the building housing the target. For example, bank robbers rely heavily on the architectural uniformity of banks. Banks often have streetcorner locations that are convenient for getaways. Glass doors permit the lookout member to see who is coming in, while those further inside have difficulty seeing through the glass because of light reflection. In their desire to create a more personal and less formal atmosphere, contemporary banks have installed low counters, giving armed robbers a further advantage.

During the reconnaissance, the group remains attentive to alarm systems, presence of guards, number of employees, and locations of safes and cash registers. The timing of the robbery is vital. In a small rural town, robbers are likely to have adequate time to carry out their crime. However, in an urban setting there is a greater risk of police intervention, as well as a larger number of customers, employees, and passersby. Many robbers plan urban crimes in great detail.

Conducting the Robbery

The procedures used during a robbery vary from one partnership to another, and from one situation or business to another. If the robbery happens in an urban area, the robbers will try hard not to draw attention to themselves, at least until the robbery is in progress. They may park their cars legally to avoid the risk of police interest, and they may wait until they are inside the business to don masks or disguises. If the targeted business is in a small town or village, robbers are unlikely to take these precautions.

Penetrating the Targeted Business. In getting to and from the bank or other targeted business, robbers often use stolen vehicles whose license plates have been interchanged with those of other vehicles. The criminals dispose of the cars after the robbery. A well-planned robbery will usually include one primary vehicle and a backup vehicle in case of mechanical failure, flat tires, or other problems.

Once inside the targeted business, the team must follow assigned roles carefully to ensure that the robbery goes smoothly. Roles are determined by the central concern of the robbery operation—*speed*. For example, one person might be assigned to watch the door and oversee the operation as it progresses, paying special attention to the time. Another may have the sole job of keeping employees and customers out of the way so that they do not interfere. In smaller teams, fulfilling this important task may mean herding employees and customers into a room or vault and keeping them under surveillance. The offenders can then focus all their attention on the owner or manager. In robbery teams of four or more, two people will usually be assigned the task of collecting the cash.

Managing Victims During the Robbery. Surprise and vulnerability are crucial elements in executing an armed robbery. Employees and customers are more vulnerable when they are caught unaware than when they have been forewarned. Surprise and the accompanying temporary paralysis of the victims allow the robbery offenders to get matters under their control quickly. Some offenders make a point of robbing businesses early in the morning, preferably on a Monday, when employees are most likely to be sleepy and dull. As the offenders see it, anything adding to surprise works in their favor.

Once they are in control of the situation, the robbers must maintain absolute authority over the ensuing tensions. The hysterical employee, the stubborn cashier, and the glory hunter may each react differently to the stress of robbery. The offenders must manage these reactions if they are to succeed at their crime. Specific techniques may vary from one situation to the next, but the main tools of the armed robbers are voice commands, a menacing appearance, and the use of force. The offenders want to sound and look as though they mean business.

Masks and hoods play an important role in establishing authority and managing tension or fear among victims. To some robbers, masks are important because they help to maintain the shock effect that was produced by the surprise entrance. To others, masks conceal not only identity but facial expressions, making it less likely for a victim to detect any anxiety or fear on their part.

Using Force.   The use of force in robbery often extends only to pushing, shoving, and on occasion hitting a victim with a gun or other object. Some robbers use an overt display of force to maintain their authority and keep the situation under control. They may attack one employee at the outset in order to convince others of their purpose and resolve and, more important, to instill cooperation through fear.

According to studies, violence arises for two main reasons:

- To help the robbers commit the crime as quickly and easily as possible
- To facilitate escape from the business after the robbery is completed

However, the same studies over many years show that professionals will rarely go beyond threats and the illusion of violence. Except in extreme situations, most professionals will not use their guns or, if they do use them, will only shoot into space rather than into people to create noise and instill fear of deadly violence. The typical armed robbery offender has no desire to seriously injure or kill. This attitude may stem from the widely shared view that those from whom offenders take money are not the real objects of attack. Robbers will hold no ill feeling against the teller, the cashier, or the customer. To most robbers, the encounter is an impersonal matter. They are not robbing a person but rather some amorphous being such as a bank, a supermarket, a liquor store, a gas station, or a loan company. The armed robbery offender is out for money, not blood. In the rare situations—a tiny percentage of business robberies each year—when an offender does use a weapon to harm a person, it is usually because the robber has a sense of being cornered or is challenged by others with weapons.

The Getaway

The getaway is the most important phase of robbery, because even if the attempt fails, the robbers need to escape. Those who have carefully laid out their operations, complete with escape routes and contingency plans, will already have their getaway well in hand.

As with the robbery itself, speed in escaping is paramount. The longer offenders remain at or near the scene of the robbery, the greater the risk of apprehension. It is during the getaway that violence such as shooting, running roadblocks, or taking hostages has the greatest chance of happening. Robbers will make every effort to clear the scene and get as far away as possible. Escaping offenders usually drive their stolen vehicles to a prearranged spot, abandon them, and then transfer to one or more vehicles owned by the team. If all goes well, the group may disperse for a time, move to a different town, or simply go underground until the "heat is off."

The Business of Armed Robbery

The popular myth depicting robbery as a senseless, violent act of plunder, perpetrated by equally senseless and violent people, can be supported on some occasions. However, in reality armed robbery is not very different from many legitimate business pursuits. It relies on such tactics as planning, organization, and management. Those who make robbery a regular pursuit (or business) will generally be involved in profitable ventures. The opportunists, or those who indulge in robbery in a repetitive but sporadic manner, will be much less profitable.

Professional robbers exhibit high levels of planning, organization, and management skills. However, opportunists out to steal a fast buck display little planning and organization. They are usually confused, fearful, and disorderly in handling their own emotions as well as their victims' stresses and tensions. Thus, opportunists present the greatest threat of bodily injury or death during a robbery, not necessarily because they intend harm at the outset but because they cannot effectively control their own fear and may strike out impulsively. Once the shooting or other violence begins, it may escalate.

Security Technique 58: Examine the Armed Robbery Risks to Your Business

First ask yourself, "What is the probability that my business will become a target for armed robbery?" On the surface, this question seems difficult to answer. Begin to assess your vulnerability by asking your local police department for information. Law enforcement departments help to collect the crime information that goes into the *Uniform Crime Report,* published by the Federal Bureau of Investigation. The *Uniform Crime Report* lists the incidence of robbery (and other crimes such as burglary) for municipalities across the nation. The report categorizes robbery by

- The total number of occurrences
- The number of occurrences per 1000 residents of the area

Most robberies involve cash-rich businesses or those selling items that can be easily converted to cash (such as diamonds, watches, and small electronic items). Therefore, your vulnerability can be determined by the nature of your business. If you sell insurance, the threat you face from armed robbery of your assets is minimal. Still, an opportunist may attempt to steal money from employees and customers in the business at the time. Other factors such as your location also affect vulnerability. For example, if your insurance company is in an area where mugging or "street crime" prevails, your risk of business robbery increases. The target is not business money, but again whatever money or jewelry can be realized from employees and customers in your business at the time of the holdup.

By contrast, if your business deals largely in cash, your vulnerability increases significantly. Although small branch banks continue to experience frequent robberies in some areas, the high risk of breaching bank security has dampened their appeal to armed robbery offenders. Professionals focus mainly on retail businesses that handle and maintain sizable amounts of cash. Supermarkets, fast-food restaurants, liquor stores, gas stations, and convenience stores are the most frequent targets. Always remember that armed robbery offenders, working alone or in groups, seek fast money and profits commensurate with the risks they take.

The *Uniform Crime Report,* mentioned earlier, also describes local and national trends in armed robberies that can help you assess whether your business vulnerability has increased or decreased. In your vulnerability assessment, you should include the local police or other law enforcement agency's record of protection. For example, in suburban or rural areas an attentive, well-organized, well-financed, and well-trained police department will have established a reputation that sends potential armed robbery offenders elsewhere for lower-risk possibilities.

Metropolitan areas pose other problems. Even if a metropolitan police department has a sufficient number of officers and excellent training and organization, its proficiency will always be limited by traffic problems, number of crimes, and other conditions that require responses daily: domestic disputes, vandalism, and a variety of mundane, noncrime-related complaints. The more populated areas create the greatest threat to a business. However, lesser populated communities that do not have established law enforcement capability businesses also are at higher risk.

Throughout your vulnerability assessment, you need to keep in mind that no business, regardless of its nature, has total immunity to armed robbery. Effective countermeasures are essential in any business, but the amount and sophistication of these measures will depend largely on the variables discussed here.

Security Technique 59: Analyze the Location and Visibility of Your Business

Your next step is to assess your business location. Do your business premises allow a potential armed robber to believe that he or she can escape easily? In a busy urban setting, with sidewalks crowded with pedestrians, an offender (especially an opportunist, addict, or alcoholic) can come into your business, get fast cash, and leave successfully by acting calmly and melting into the sea of people. The offender may then board a bus or train, catch a taxi, or meet an accomplice nearby with a car that blends into the traffic.

Regardless of location, you need to remember the discussion of visibility in Chap. 1—in particular, avoid concealing your business interior by blocking windows with advertising posters. You also need to examine the location of your cash. Consider keeping it in a place where passersby can easily observe an ongoing robbery. Visibility might be the greatest deterrent, because even the most desperate or careless armed robbers will realize that their chances of committing the crime and successfully escaping are slim. Your best protection is openness: an area free of any possible means of concealment. Offenders will have to commit their crimes in full view of those looking in, and few will ever take that risk.

Assessing Entry and Escape Routes

So far, you have determined the threat level to your business from the *Uniform Crime Report*, increased your business visibility, and situated your money up front where those outside can easily see an offender attempting to rob you. Now consider how the criminal can get in and out. The front door is the most obvious way, but other possibilities include the roof, back doors, and ventilation ducts—or even hiding as a "stay behind" after the business closes. You need to assess these possibilities, even though most armed robbers are not that ambitious or inventive. Assuming a robbery will occur, put yourself in the place of an offender who decides to be a bit unconventional and not use the front door. How would you do it? You need to be creative, and look for any possible entry and exit.

Security Technique 60: Determine How to Respond to Armed Robbery

Once a robbery is in progress, is there any way you can gain control or, in other terms, remove control of the situation from the offender? Remember that the armed robbery offender relies on maintaining control. In some situations, business owners have pulled out a gun and had a shootout with a robbery offender. However, unless you have law enforcement training or ideal confrontation conditions, you should never attempt that kind of action.

Assuming you are not qualified to mount a successful confrontation, consider that your best course of action is to let the police know what's happening. Opening up your business to view from the outside is the first part of this effort. However, the robbery may happen when there are no people outside looking in. You may be alone with the armed offender and need to summon help. Duress alarms provide a good opportunity to take that action. Examples of duress systems are given in Chap. 4, and they work effectively when placed where someone in the business can activate them. Don't make the mistake of having duress buttons in only one location. Also, make sure that employees know where duress buttons are and how they work.

In a metropolitan area, a duress system will alert a central security company, which notifies the police. In smaller communities, the police may agree to have the alarm installed directly in their dispatch room—and that is your best protection. An inexpensive backup system can also help, especially when the duress alarm cannot be linked to a security company or police department. A series of bright flashing lights (strobe lights) fixed to the front, rear, and sides of your roof, and connected to the duress buttons, can summon help. However, you need to ensure that local law enforcement officers know what these flashing lights mean. Also, if any other businesses or residents are within sight of the system, be sure they know to call the police on your behalf when the lights flash.

In some areas, you might consider adding a loud external bell or siren that will be clearly audible inside as well. Once set off by a duress button, it attracts attention and draws police. It will certainly have an important impact on the offender, who will quickly sense a loss of concealment as well as control. In one southern state, a supermarket that was a target of armed robbers about twice a year added all these countermeasures—plus one more. It wired the duress system so that the security company received notification, lights on the roof flashed, and a loud piercing siren on the roof sounded. In addition, a series of bright interior lights began flashing. According to the manager, when a group entered the store to commit a robbery and all systems were activated (by one employee pushing one button), the robbers became so disoriented and afraid that they ran. They were soon arrested by the police. Since that time, no robberies have occurred at that supermarket.

Your protection needs may not call for such an elaborate scheme. But keep in mind that any overt countermeasure takes control away from the offenders by "blowing" their concealment and making escape doubtful if they remain in the area. The absence of the three important conditions for a successful robbery—control, concealment, and escape—makes *your* business too high a risk. Through the elimination of opportunity you can and will eliminate the crime.

Security Technique 61: Learn to Recognize an Imminent Robbery Threat

Another important countermeasure that removes control from armed robbers is to recognize them before they have the opportunity to surprise you. Remember that surprise is one of the keys to success in armed robbery. You can reduce surprise by maintaining close liaison with your local or metropolitan police department (in the latter case, with the precinct that has responsibility for your area). I suggest that at least once each month you go to the police and inquire about any new robbery trends that have emerged. Often a quasi-professional or young (even juvenile) group will enjoy so much success at robbery that it is encouraged to continue. Or a professional group will work a series of nearby towns or small cities and then move across country to resume operations. This type of professional group depends on reaping as much profit as possible, and then disappearing from the area when the pressure for arrest builds up. When you keep in touch with law enforcement officials, you can add or decrease countermeasures to your asset protection systems according to their reports.

Keep a log or newspaper file of robberies targeting businesses in and around your premises. News clippings will often alert you to trends, and offer "glimpses" of physical descriptions, operational methods, personality types, and other valuable information. A shot from a newspaper could enable you to identify a person lurking around your business or posing as a customer. This person might display more than a passing interest in your operations. Ensure that you keep your employees informed, because the more eyes you have watching for suspicious people (especially those who fit the composite descriptions supplied by the police or newspapers), the more likely you are to prevent a robbery or at least to ensure that the criminals are apprehended.

When you openly support your local law enforcement officials and maintain a working liaison with them, you can ask for their help if you believe a suspicious person is in your place of business. The police might arrive to thwart a robbery or to identify someone wanted for robbery or some other crime. Once a person has been identified—even if he or she hasn't committed a crime—the person is unlikely to threaten your business location again.

Security Technique 62: Use Closed-Circuit Television

Another good robbery deterrent is to attach closed-circuit television systems to a secured video recorder. Like the systems used in banks, the video arrangement allows you to watch the actions of those in your business premises without making it apparent that you are doing so. It's a good idea to record a clear picture of suspicious people through a remote switch. Preserve the tapes, ask the police to view them, and keep the tapes for reference in case you later suspect that the same people committed a robbery. Often, even when masked, they will have certain identifiable characteristics such as rings, clothing, and tattoos. However, a closed-circuit television system will often be enough, in itself, to thwart an attempt at armed robbery.

Security Technique 63: Observe Suspects Carefully

More often than not, an armed robbery offender will enter the business and look over the situation before attempting the robbery, or will return later to commit the crime. Young professionals, amateurs, and opportunist robbers will regularly give themselves away by their actions, and you need to remain alert to them. When "casing" the business, or weighing the possibilities of success, inexperienced offenders will appear too concerned about trying to blend in with the other customers. They will often behave as if someone on the store's staff is watching them. In reality, this may not be the case, but such actions will make would-be offenders stand out from the crowd. They may become nervous when store personnel approach or may fumble around with merchandise, pretending to have an interest in purchasing. By contrast, seasoned professionals will dress like the average customer and go about the normal prepurchase actions.

When trying to identify a potential armed robbery offender, use one basic rule. Always look for the situation or person who does not fit into the normal course of your business operations. Discreetly and carefully study the person's behavior and contact the police if you feel the least amount threatened. (Appendix A provides valuable guidelines for observing suspects or offenders so that you can later help the police or testify in court proceedings.)

Security Technique 64: Stay Alert When Robbery Happens at Your Place of Business

Despite all your protection efforts, a robbery can happen at your place of business. I've read and heard all types of advice about what to do. Clearly, when the offender has control of your business, you and all others there at the time should do whatever the armed robber wants you to do. Earlier, I discussed the countermeasures that will bring you success in the prevention of robbery. However, even the most alert individuals and the best security effort cannot deter the determined idiot or the masterful professional. These criminals believe the obstacles are no match for their skill.

While doing as you're told by the armed offender, you should also consider how you can remove the control. I've shown you a variety of ways to do so. Remember that most robbers want money or items easily converted to money, and they want to get out as soon as possible. Your greatest danger is in situations where the easy money doesn't happen. Although violence is rare, sometimes offenders will injure or kill to prevent identification. Over the years I've had the task of investigating robbery situations where, despite obvious cooperation, the victims were either killed on site or transported as hostages to another location and then killed by the offenders. Don't ever assume that mere cooperation can prevent deadly violence.

I am not advocating fighting armed offenders. However, if you immediately put in place all the techniques I've noted above, remain interested and attentive, and maintain close liaison with the police, you will have a strong prevention atmosphere that is easily recognized by armed offenders. Next, you need to carefully assess whether you can remove control from the robbery offenders so they cannot steal from your business and escape. With some offenders, improper attention to this problem could cost you not only your business but your life—as well as the lives of your employees and customers. Take the suggested action today and protect all your assets from this type of crime.

Quick Reference Guide to Chapter 5: Eliminating Business Robberies

Figure 5-1 presents a quick reference guide for protecting your business against robberies. Once again, note the fundamental principle involved: If you eliminate the opportunity, you will eliminate the crime.

Four primary categories of offenders
   professional
   opportunist
   addict
   alcoholic

The anatomy of armed robbery
   planning
   organization
   management

Figure 5-1. Eliminating armed robbery at your business premises

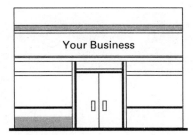

Assess your business vulnerability

Look for opportunities

Eliminate the opportunities

Maintain close liaison with police

Use the uniform crime report
and newspaper accounts of armed
robbery to create advantage

Eliminate the opportunity
Eliminate the crime.

Take back control of your
business using a variety of
devices and techniques.

Figure 5-1. (*Continued*)

6
Putting an End to Employee Pilferage

Employee pilferage causes more businesses to close or go into bankruptcy than any other crime. Studies by the government and by a variety of business organizations—and my own experience helping businesses—indicate that employee pilferage accounts for 38–75 percent of business losses. I need to add that business owners and management often resist the idea that employees steal from them. Others know or suspect it but ignore the problem, believing it's only a temporary situation. I call this the "ostrich approach" to theft.

The retail industry, for example, estimates that from 75–80 percent of its inventory shrinkages results from various types of employee dishonesty. A study by the U.S. Department of Commerce shows that employee theft in manufacturing plants totals about *$8 million a day nationwide.* Employees, however, rarely view their theft as having such a major impact on the business that employs them. They create a variety of justifications and rationalizations for their acts:

"The company doesn't mind."

"They've got plenty, and they're not losing anything on what I take."

"Everyone is doing it."

"They don't pay me what I'm worth so I'll just take it in other ways."

"They don't pay me much and they expect me to take it in other ways."

112

Unfortunately, most employers know pilferage occurs but rarely do anything about it. Believing that their losses stem from something other than employee pilferage, some employers impose severe or oppressive controls. These actions often increase employee theft because the business owner is convinced the loss comes from some other activity. Like any crime that targets your business, you don't have to allow it to continue. You can end employee pilferage. In this chapter I'll show you tested ways of doing so without interfering with business operations or creating a nightmare of stringent controls.

Understanding the Magnitude of Employee Pilferage

No business, despite its size, has immunity from the devastating crime of employee pilferage. Some years ago I heard a story about Joe, a working man who had been employed for 30 years by a large diamond mine. At the end of each shift, company security guards would carefully search each worker leaving the mine. They often found someone trying to smuggle out a diamond. Joe, however, was known for his honesty, and for three decades he appeared at shift's end pushing a wheelbarrow and openly submitting to the security search.

After retiring from the company, Joe was not heard from again. Then one year the diamond company's security director retired and took his wife to a tropical island for a vacation. While walking around the island paradise, they came upon a palatial mansion and noticed a man sitting on a spacious veranda. As they walked by, the security director recognized the man as a former worker at the mine. The director waved and shouted a greeting to Joe, who recognized him and invited them to join him on the veranda. After exchanging small talk, the retired security director asked Joe how he could possibly afford such an estate. Joe smiled and said, "I made my money from saving hard and from selling wheelbarrows."

The security director nodded but seemed to miss the true meaning of Joe's statement.

During his employment at the diamond mine, Joe took home, not diamonds, but, a wheelbarrow each day for 5 days a week. At home, he cleaned the implement, repainted it, and sold it to the company that sold wheelbarrows to the diamond mine. He stole about 7500 wheelbarrows during his tenure, for about $100 each. The supplier sold them back to the mine for $150. Over a 30-year period Joe, the hardworking commoner valued for loyalty and honesty, made about $750,000. Adding his personal savings of over $100,000 from frugal living, plus accrued interest, he was able to retire with well over $1 million, a comfortable sum for living on a tropical island.

I use this example often when talking to business owners in seminars, or to clients, to create an awareness of how easily an employee can siphon business profits without appearing to do so. For example, at one client's company several hundred employees regularly took office supplies, tools, and small equipment home. Although realizing that operating costs were excessive, the owner resisted the idea of the employees' stealing company property. A complete inventory—coupled with a detailed study of what the company's average operational expense should be compared with its business volume—revealed a staggering loss of over 50 percent of supplies purchased. The problem was solved easily by establishing a control system. After a year, with the controls in place, the company spent 55 percent less than in previous years.

Over the years, I have encountered similar problems in a wide variety of businesses, including restaurants, warehouses, and retail stores. The solutions supplied in this chapter are the ones that worked effectively for these businesses.

Security Technique 65: Recognize Employee Pilferage at Retail Stores

Before you can establish controls that end employee pilferage, you need to recognize the problem and develop an understanding of how it happens. Here are some examples from retail stores.

Problem 1. Employees may remove shelf items, replacement stock, and other supplies when the retail outlet closes to patrons. Such pilferage often occurs at night or at other times when merchandise is restocked. Stolen property may be taken directly to personal vehicles or hidden in other locations. Employees may also hide business assets in empty cartons, garbage cans, or other refuse containers for later removal from the store, either with or without the knowledge and assistance of other employees. Such pilferage may include both perishable and nonperishable items.

Problem 2. Shelf items or merchandise normally kept in closed display cases, usually small and frequently of considerable value, may be hidden directly on the employee's person, in clothing, or in pocketbooks. Items may be placed in shopping bags containing purchases that were made earlier. Employees may pilfer from areas in the store where employees are not assigned or accountable but to which they have uncontrolled access. Failure to inspect employee purchases and permitting employees to purchase at nonscheduled times and from other than designated personnel provide opportunities for this type of theft.

Problem 3. Consumption pilferage, either systematic or casual, is an often overlooked or ignored method of employee theft. Items such as tobacco, edibles, beverages, clothing, cosmetics, and other personal items considered by some employees to be "fringe benefits" are surreptitiously stolen and used or consumed in areas of the store or other parts of your business facility. Employees often retain these items or the unused parts of them for future use and keep them on their persons, in pocketbooks, or on shelves, under counters, or in various other work areas.

Problem 4. Collusion may exist between vendors and your employees. For example, an employee signs for more merchandise than was delivered, receiving cash or merchandise from the vendor as remuneration. This method becomes particularly attractive to the dishonest employee when the commercial supplier personally fills consumer outlet shelves or stores merchandise for later shelf stocking by employees. The dishonest employee can stay remote from the vendor, the merchandise, and the mechanics of storage or stocking, and it will appear that he or she is uninvolved in theft.

Problem 5. A comparable pilferage-and-remuneration scheme may exist between employees and garbage or trash disposal personnel. In this collusion system, various goods such as food items are collected as garbage and then removed from trash containers for later personal use, or for sale to others for cash.

Security Technique 66: Examine Indirect Employee Theft for Financial Gain

Indirect theft can be as costly to businesses as outright pilferage. Employees sometimes use unauthorized privileges or assist in pilferage to realize financial gain for themselves or their customers.

Problem 1. Employees may indirectly engage in theft by marking down valid merchandise prices for favored patrons or other employees. Techniques include stamping a reduced price on an item and changing or switching price tags.

Problem 2. Employees may fraudulently declare merchandise old, shopworn, damaged, or salvaged in order to create a reason for reduced prices. Schemes include falsely reporting commodity condition, breaking or bending merchandise or containers, cutting seams on clothing, and deliberately smudging or otherwise changing the external appearance of merchandise without materially altering its usefulness.

Problem 3. Employees may favor their friends with extra amounts of merchandise, including the foods and beverages that they serve. Careless waste of foods and other perishable merchandise or expendable supplies amounts to a substantial annual loss. (Deliberate waste is common among disgruntled employees trying to get back at management.)

Security Technique 67:
Recognize Other Methods of
Employee Theft

Several methods of employee pilferage involve the use of the cash register at patron checkout and loading areas. These techniques can lead to frequent and high losses.

Problem 1. The cashier or other employees may steal cash directly from the register.

Problem 2. After hours, or while unobserved, employees may rerun the register tape at lower figures, clear the register with the lowered total indicating actual receipts for the day, and then pocket the difference.

Problem 3. Employees may pilfer items by attaching a valid register receipt, obtained from a patron purchase, to a bag or package that contains (or will later contain) stolen business assets. The employee may place the open bag in a wastebasket to accumulate pilfered items during the day. After the business closes, the employee staples the acquired cash register tape to the bag and carries it out of the store as a valid purchase made earlier.

Problem 4. The employee reports false overrings, receives credit for them, and pilfers that amount as cash.

Problem 5. The employee reports refunds for fictitious merchandise allegedly returned by dissatisfied customers. The employee then steals the refund amount by falsifying the documentation and signatures required for valid refunds to patrons.

The false-refund scheme may be enhanced by collusion between an employee and a patron. The patron signs falsified documentation supplied by an employee alleging that merchandise was returned. The two then split the refund for the nonexistent merchandise.

Problem 6. "Carryout" employees remove items from bags or containers following valid patron payment. Typically, the pilferage occurs between the time the patron leaves the checkout area and the time he or she returns with a vehicle. The technique may involve both cashiers and baggers, either individually or in collusion.

Problem 7. Pieceworkers zip on finished garments as personal clothing and take them home.

Problem 8. Cashiers ring up lower prices on purchases and pocket the difference. Some work with a customer accomplice and ring up low prices as the accomplice passes through the checkout line.

You might think that modern technology—such as price code scanners, computerized cash register totals, and programs calling for entry of code numbers—would put an end to this type of theft. Unfortunately, whatever system you use may stop one type of pilferage but facilitate another. For example, if an accomplice enters your business and buys five items, there is nothing to prevent the cashier from ringing up only two. Or suppose the cashier has accumulated a $50 excess in the cash drawer. When the accomplice buys a small item, the cashier pays out the $50 in change. With very large amounts, the cashier may use two or three accomplices to dispose of the excess money.

Problem 9. Receiving clerks obtain duplicate keys to storage facilities and return after hours to pilfer items. They may alter bills of lading and invoices to make them appear as short shipments or backorder items.

Problem 10. Truck drivers make fictitious purchases of fuel and repairs and then split the gains with truck stop owners.

Problem 11. In cooperation with a receiving employee at a distributorship, a trucker keeps a certain percentage of the goods delivered. For example, the store employee signs a bill of lading for 100 cases and the trucker keeps 20. The two split the profits after the driver sells the stolen goods to another store or distributor that is not particular about its sources. Or the trucker may sell the pilfered goods to roadside buyers from the tailgate of the truck.

Problem 12. An employee copies computerized business information (either on disk or by printout) and sells it to a competitor. The competitor then steals your business by undercutting prices, mounting a sale before you do, or taking advantage when your business is financially vulnerable.

In one situation, two retail stores competed briskly for 2 years. An employee of Store A copied computerized business information onto disks and sold it to Store B for cash. The employee also was promised a better-paying job at a later date at the competitor's site. The information showed that Store A had a tight cash problem, and the manager of Store B discreetly mentioned to vendors that the competition might be in financial trouble. The innuendo was enough to stop further credit lines. Vendors began asking Store A for payment of all accounts, and future shipments were to be sent COD. Unable to restock and pressured from its accounts payable, Store A went broke within weeks.

Problem 13. Fast-food or other restaurant employees eat costly foodstuffs while they work. Some put items into handbags or containers to take home with them at the end of the day. Others steal a ham or steaks, even silverware and utensils, by tossing them in the trash and recovering them after work.

I once explored a problem of huge losses at a high-volume store owned by a major fast-food chain (a name you would readily recognize) in the South. The store did enormous business but consistently lost money. My investigation disclosed widescale pilferage—involving everyone from managers and cashiers to maintenance workers. The store had over 90 employees, and each took some cash or product every day. In addition, employees ate a large amount of finished product. Management was changed and a few reasonable controls, policies, and procedures were introduced. The losses quickly stopped.

Security Technique 68:
Recognize Employee Pilferage
in Warehouse Operations

Employee pilferage is commonplace in warehouse operations—limited only by opportunity and by existing operational controls and physical security measures. Here are some typical methods employees use to steal from a warehouse operation.

Problem 1. Employees remove stock after gaining access to service entrance keys or to unauthorized duplicate keys.

Problem 2. Warehouse doors, windows, and other openings are deliberately left unlocked. Employees intentionally hide or arrange to be locked inside the warehouse after the workday (as "stay behinds") to pilfer stock from the premises.

Problem 3. An employee is deliberately careless or inaccurate in issuing or receiving warehouse stock. The resulting surplus, for which the employee is not accountable, is then removed for personal use or for sale to others.

Problem 4. Employees remove items from cartons or other containers, reseal the pilfered containers, and return them to their proper location.

Problem 5. Employees alter stock records to cover up shortages of items stolen earlier or items intended for pilferage. Carriers and warehouse personnel may act in collusion in this type of scheme. Stock scheduled for systematic pilferage may already be in the warehouse, or may be en route or pending delivery, and typically includes clothing, food, automotive items, hardware, electrical appliances, and other high-value low-bulk items.

Problem 6. Employees deliberately falsify internal inventories, issuance and receipt records, and other accountability documentation. The business is especially vulnerable when inventories and audits are conducted by other than honest, qualified, or disinterested personnel.

Security Technique 69: Assess Employee Pilferage in Manufacturing Plants

A manufacturing plant supplies a treasure trove of opportunities for employee pilferage. Even the most common items in a manufacturing plant—a handful of nuts and bolts—are ready candidates for theft. Employees can easily snip off several feet of wire for a home repair job or pocket a small hand tool.

One manufacturing client called me in to find viable solutions to loss problems. I found that employees had stolen $12,000 worth of brooms and $57,000 worth of cardboard shipping containers in a single year. Besides these common items, employees had pilfered about $150,000 worth of hand tools, components, and other items from the medium-size plant. The business was fast approaching bankruptcy. I was able to turn the situation around and end the pilferage by immediately instituting a series of controls, beginning with strict and systematic inventory systems.

Here are other common examples of plant pilferage:

- Employee theft of scrap metals amounted to losses of over $250,000 in 2 years.

- Employee theft of component parts and tools created losses of over $1 million in 1 year.

- Over a single weekend, employees in league with outsiders stole $300,000 worth of finished products on pallets ready for shipment.

- A fabric manufacturer lost $350,000 worth of cloth from employee pilferage in 2 years.

All these problems can be (and were) stopped cold by the assessment techniques and inventory controls described later in the chapter.

Security Technique 70: Evaluate Threats of Employee Theft in an Office Environment

When a business operates exclusively from offices, it is difficult to understand how employees can steal very much without their thefts becoming readily apparent to employers. For example, a typewriter or computer system in regular use would be quickly missed. Also, such items are large enough to make concealment difficult.

Over the years I've had many clients who are baffled at year's end when their operating costs are high and the disappearance of equipment cannot be explained. In each case, besides a lack of daily accountability, I discovered that employees were afforded accessibility to offices at night and on weekends. Even buildings that had 24-hour security officers at entrances and exits, and that required coded entry cards showing the time, date, and identity of employees coming and going during nonbusiness hours, experienced severe loss problems. The employers were convinced that no one could steal their business assets under those stringent conditions. The problem is that tight security focuses on *unauthorized intruders—not on authorized employees* entering and leaving.

Most employee pilferage in an office environment occurs under the guise of legitimate business operations, and much involves theft of administrative items viewed as "expendable." Employees have put their children through school on pencils, paper, sharpeners, staplers, tape, and other supplies stolen from an employer. They simply carry out a few items each day in a pocket, a purse, or a briefcase. In one case, an employee sent out the items through the business mailroom. Small thefts soon add up, especially when nearly all employees take part, even though each acts independently.

Another client of mine was confronted with an ingenious method of employee pilferage. The client maintained a great deal of office equipment, including desktop and portable computers, typewriters of various types, calculators, copiers, fax machines, and video equipment. "Normal" pilferage took the form of employees mailing out videotapes to relatives (or to their own post office boxes) under the guise of sending these items to legitimate customers.

However, two employees had a grander scheme. They ordered large quantities of administrative supplies by mail and had them delivered to an "annex" address that was the one-person office of a relative. Their orders included computer printer ribbons and other items that were not compatible with the employer's operations—but no one noticed. In another instance, they managed to smuggle out typewriters, printers, and computers by having a friend impersonate a service representative from a computer and office equipment shop who was "picking up" items for repair.

The employees were not held accountable for the missing items. They destroyed or altered file copies of invoices and other records. Ironically, even though the annual audit showed large business losses, the auditors could not pinpoint the losses or determine why operating expenses were high.

Why Do Employees Steal?

Once you understand the magnitude of employee pilferage and your business's vulnerability to it, you need to dissect the crime in detail. What prompts your employees to steal? How can you anticipate their activities and reach to the core of their motives? What is the typical pilferer's "profile" and how can you turn the pilferer's methods to your advantage? The following techniques help you answer these questions as you work to make your business immune to employee theft.

Security Technique 71:
Anticipate Employee Pilferage_____

Employees who steal business assets rarely accept a job for the purpose of theft. However, financial pressures, lifestyle aspirations, or other perceived needs may encourage them to pilfer. Once it starts, pilferage becomes an addiction. The pilferer is like a person with increased income who finds it hard to forfeit a comfortable lifestyle once it is in place.

Although we don't normally think of pilferage as an additional source of income, it is just that. If an employee steals food, he or she will not have to use money to buy food items. If the employee steals clothing, tools, or office supplies out of a perceived or genuine need, the employee saves money by not purchasing the items. The greater the perceived need, the greater the likelihood that an employee will pilfer your business assets.

Employee theft extends to all levels of the organizational structure. Corporate executives who make large incomes may steal corporate assets because their lifestyles have become greater than their incomes will support. A clerical worker might use the business copier to duplicate flyers for a sideline business, send out resumes over the fax machine, or make long-distance calls for personal reasons. Other employees may pilfer tools or equipment and supplies and stockpile them in the garage until they have the opportunity to start their own business. Whatever their reasons for stealing, all employees can easily become pilferers, especially when you supply them with the most important factor: opportunity.

I would *like* to report that most employees will not steal your business assets. However, my personal experiences and those of scores of loss prevention managers and bonding companies—along with countless studies and the endless statistics on bankrupt businesses across the country—indicate that this is not the case. The hard truth is that people will be dishonest when given the opportunity. Most employees will steal if they get a chance, a sizable number will steal if they think they can get away with it, and only a precious, small minority will not steal under any circumstances.

Security Technique 72: Assemble a Profile of Pilferers

There are two categories of employee pilferers you need to recognize and counteract in your business: those who steal casually and those who steal systematically.

The Casual Pilferer

The casual pilferer steals primarily because he or she cannot resist the temptation of an unexpected opportunity. The employee has little or no fear of detection and usually acts alone. Casual pilferers steal without premeditation or planning and may even take items for which they have no immediate or foreseeable use. Typically, they steal small quantities of supplies for family or friends, or for use around the home.

Casual pilferage happens whenever an employee feels the need or desire for a certain article and the opportunity to take it arises because of poor asset protection measures. Although casual pilferage involves unsystematic theft of small articles, it is a serious crime. Major losses can result if you permit it to become widespread in your business, especially if the stolen items have high cash value. Further, there is always the possibility that casual pilferers, encouraged by their success, may turn to systematic theft.

The Systematic Pilferer

The systematic pilferer steals business assets according to a preconceived plan, and steals all types of supplies or merchandise to sell for cash or to barter for other valuable or desirable commodities. Occasionally, these pilferers also steal items for personal use.

Systematic pilferers may work in pairs or groups, or may operate in collusion with business outsiders. They may be members of a cleaning crew or even well-placed administrative workers who have direct control over the desired items. Their thefts may occur from storage areas or transit facilities.

Pilferage can be a one-time event or a habitual, career-long activity. Large quantities of supplies, with significant value, may easily be lost to groups of employees who elaborately plan and carefully execute pilferage schemes.

*Security Technique 73: Understand the Motivation of Employee Pilferers*_____

The uses that pilferers make of stolen items (or the money obtained from selling them) do not fall into universal patterns. Their methods of operation are equally difficult to pin down because they vary with individual motivation. Here are the common danger signs that a pilferer is at work in your business:

- Dedication or devotion to work, especially to mundane jobs, that exceeds normal interest
- Increase in personal spending
- Refusal to accept promotions or transfers

Pilferers are encouraged to ply their trade by (1) opportunity, (2) high personal need or desire, and (3) ready rationalizations for their actions. Typical rationalizations of employee dishonesty include:

"Why not? Others do it."

"It's morally right for me."

"It's not stealing, only borrowing."

"I'm worth more than I'm paid and the boss expects me to steal."

Security Technique 74: Know Your Business's Vulnerability to Pilferage

How vulnerable is your business to pilfering? In making your assessment, keep in mind that any business has pilferable items, products, supplies, or merchandise. Often you can determine if you have an ongoing problem by analyzing what your operations *should* cost, and what your net profit *should* be, compared with what it is. I use the term "net profit" because that is the business asset siphoned by pilferage. It is what you should have after deducting all legitimate expenses (insurance, utilities, and other operating costs) from "gross profit." Also, you need to determine the various categories of operating items and decide what normal business operations would consume as compared with actual consumption of these items.

A greater sense of urgency arises if a problem already exists and you need to focus on controlling those areas that have been identified. This process will supply you with the identity of an employee, or at least a group of employees, responsible for the problem.

Put yourself in the place of an employee who decides to steal business assets. Adopt a creative approach and look for all possible ways to beat the controls or systems in place. Chances are that when you assess your business in this manner, you will be shocked by the ease with which employees can steal.

Security Technique 75: Identify Top Targets for Employee Pilferage

Both casual and systematic pilferers have certain problems to overcome and requirements to fulfill when stealing your business assets. Use the following guidelines to identify the top targets in your business for employee pilferage.

Would-be pilferers must first find the item or business asset they want to steal. With casual pilferage, the discovery may happen by accident or by a personal search of assets. With systematic pilferage, more extensive means are used to plan the theft.

Pilferers must then determine the manner in which they can gain access to, and possession of, the desired items. The method may be as simple as breaking open a box. Or it may be as complex as surveying your business for weaknesses in physical safeguards, checking your security procedures for loopholes, or forging shipping documents.

Next, pilferers must find a way to remove the stolen items. Some may wear special articles of clothing that provide for concealment of small items; others may rely on vehicles to remove larger business assets from the premises. Through falsification of documents, employees can arrange for removal of whole truckloads of supplies without immediate discovery.

Finally, to obtain any benefit from their crime, pilferers must use the stolen business asset or dispose of it in some way. Casual pilferers will use the asset to satisfy a personal desire or need, or perception of a need. Systematic pilferers usually sell stolen items through fences, pawnbrokers, or black marketeers, or keep the items for their own or a friend's sideline business.

Security Technique 76: Use the Pilferer's Methods to Your Advantage

Systematic pilferers choose a target for its monetary value, since their goal is personal profit. They seek out items that will yield the greatest financial gain. This means they must also have a ready market for the business assets that they steal. These pilferers look for small items of high value—drugs, precious metals, tools, or electronic devices—in their place of business. However, if the profit is substantial, systematic pilferers may also select a target of great size and weight. Bulk storage areas contain most of the materials that will satisfy their needs.

Casual pilferers are likely to take any item that is easily accessible to them. Since they remove the item from the business by concealing it on their person or in an automobile, size also becomes an important consideration. Bulky or heavy items will not be on their agenda. Nor are monetary value and available markets a concern of casual pilferers, since these employees usually do not plan to sell the stolen property.

Your "pilferage prevention" program should begin with a careful study of these employee techniques. Pay particular attention to how stolen items are used or disposed of. Each line of attack from pilferers can become a line of defense for your business.

Inventory Accountability and Control

One of the most neglected aspects of business operations is inventory control. Some businesses invest great effort in certain inventory areas but neglect others. They especially neglect expendable categories. Employees who decide to pilfer business assets will quickly determine what you safeguard and what you don't. If your existing inventory system does not mandate accountability for every item on your premises, it presents an open opportunity for employee pilferage.

Assume your business includes selling soft drinks or coffee by the cup, and that this operation is booming. The profit margins for this element of your business remain high, as does the net profit. However, at the close of the month, your resupply bills, payroll, utilities, and other expenses show that you lost money on the sales. You will probably suspect that either sales were not recorded accurately or your employees were stealing. Most businesses tend to stop investigating there, and the problem continues. A few businesses may try to set traps to catch a suspected employee—and sometimes they do. However, catching one dishonest employee doesn't "set an example" for eliminating theft. Other employees are likely to continue their pilferage. Most important, the *opportunity* to steal remains unchecked.

The best way to eliminate opportunity is to take aboveboard action, and that begins with a daily inventory in some situations and regular inventories in others. The process may sound time-consuming—one more task to fit into an already busy schedule. However, since you are in business to make profits, you cannot afford *not* to take the time. Also, when you develop a reliable system of controls, the inventory process moves more quickly and reaps large benefits.

The most important advice I can give you is not to wait for an accountant or auditor to tell you that your losses exceed your profits. If you delay taking action—and unfortunately most business owners do—it might be too late to save your business. Start controlling your assets today, and do it with a reliable and routine inventory system.

Security Technique 77: Create a Viable, Tailored Inventory System

When I mention daily inventories as a control measure to clients, most look at me as if I came from Mars. I know all the reasons that you should not or cannot conduct a daily inventory. However, my argument remains the same. Do you want to make money and have sound profits from your investment and efforts? If the answer is yes, then you have to begin an inventory system today!

Setting up a system is not difficult or time-consuming, whether your business is an office, retail store, warehouse, manufacturing plant, or other enterprise. You don't have to inventory every item in the business—only those that are easily stolen. In fact, if you count your daily receipts at the close of each business day and compare them with the register tape total, overrings, and cash payouts, you have already established 50 percent of the daily inventory system that I suggest. However, knowing what money you acquired throughout the business day does not tell you whether that amount is correct.

Daily Inventory Reports and Worksheets

My wife and I are business owners. Our sons manage the operations for two restaurants and a retail store. I manage the assets. Figure 6-1 shows the daily income report that each of our cashiers must complete at the end of a shift. The report is an important part of the daily inventory process at all our locations.

Figure 6-2 shows the daily inventory worksheet we use. Notice that each worksheet is filled out by the department employee and then verified by a manager (and sometimes by a member of our family). By using these two forms, I can maintain controls that prevent employees from pilfering cash, merchandise, or food without being immediately detected.

According to the well-known adage, unless you know where you are now, you cannot know where you are going. This precept is equally true of the systems you need to establish to eliminate pilferage opportunities and to detect thefts as they occur. You need to establish an initial total inventory as your starting point. Once you learn exactly where you are, the daily income report combined with departmental inventories will tell you exactly where you're going and where you will stand at that time. As you reorder, the changes will be reflected in the inventory worksheet, and it's a good idea to make a marginal note of the incoming shipment to verify with the invoice.

Checks and Balances: The Layered Approach

To begin a viable business asset count system, first look at your business in sections, not as a whole. For example, in an office environment, you would want to inventory equipment and supplies; in a retail environment, you should inventory the stockroom and specific display and

Bintliff's

Date: _____   Register No: _____

Shift: _____   Cashier: _____

Cashier's Closing Count of Business Income

Ones		Pennies		Checks		Chg Accts	
Fives		Nickels		C. Cards			
Tens		Dimes		T. Exch			
Twenties		Quarters				Gift Cert. Sales	
Fifties		Others				Gift Cert. Redeemed	
Hundreds							
TOTAL		TOTAL		TOTAL		TOTAL	

Collective Total Columns 1 – 3 [   ]

Reconciliation

		Staple Cash Register Tape Here
Register Z Total		Notes
Starting Bank *		
Subtotal		
Overrings *		
Subtotal		
Paid Out *		
Balance Total		
Total Drawer Count (at close of each shift) *		
Differences Plus or Minus		

Cashiers: Fill in only the blocks with an asterisk* in the reconciliation column above

Figure 6-1. Daily income report.

sales areas. If your business involves a warehouse or manufacturing plant, decide which of the assets (tools, supplies, information, documents, and so on) will supply the greatest opportunity to pilfer.

Some restaurants regularly use serial-numbered order receipts in the belief that it solves the problem of employee theft. However, this tight control system does not totally eliminate the opportunity to steal. Employees can steal in other ways, or can work with an accomplice-customer to undercharge on checks. The tracking of receipts establishes only one layer of inventory management.

Bintliff's

Inventory Control Worksheet

Date: _____   Dept: _____

Person Conducting Inventory: _____

Item Description	On Floor/In Use	Storage

Notes:

Figure 6-2. Daily inventory worksheet.

An effective inventory control system, tailored to your specific business operations, will have several checks and balances. Making your employees a part of the accountability process supplies an important benefit. It acts as an effective deterrent against pilferage by heightening the employees' awareness of effective controls. Remember that the pilferer relies on stealing without the business owner detecting the shortage until much later, and further relies on not becoming identifiable as a thief. Your objective should always remain the same: eliminating the opportunity, rather than "catching" the thief in the act. If your inventory system and other controls remain effective and constant, pilferage of your business assets will come to a virtual halt.

Security Technique 78: Adapt Your Joint Inventory to Other Business Needs

The concepts shown in Figs. 6-1 and 6-2 are easily adapted to a warehouse, office, manufacturing plant, or other type of business. The important factor I stress to clients is not the specific forms they create, but the systematic controls that daily or routine inventories provide.

The system also acts as a deterrent to employee pilferage. Remember that aside from the occasional systematic pilferer who intends to sell stolen goods for personal gain, most employees steal because the workplace provides a tempting opportunity. Routine inventories and accountability will eliminate the opportunities for all types of pilferage.

Again, involve your employees in the routine inventory procedure. Have the employee responsible for each section of the business conduct a close-of-business inventory. This creates a sense of accountability and responsibility, making the employee feel a part of the business, and also largely removes the temptation of pilferage. Verify the employee's count discreetly, to ensure its accuracy. My experience as a business owner and security consultant suggests that employees will rarely make a deliberate inaccurate inventory, because they assume it will or may be verified by management.

When you tightly control items directly involved in your business, employees who are inclined to steal may look for other avenues, and that often leads them to expendable items. In an office environment, these items normally encompass administrative supplies (paper, pencils, clips, and so on) plus small equipment such as calculators. In a manufacturing and warehouse environment, the expendables might include tools, oil, gloves, and other hardware items. A retail store can have a variety of display aids, administrative supplies, tools, and other items that are easily stolen without accountability. A restaurant or fast-food facility contains edibles and other small items that casual pilferers can use at home or that systematic pilferers might use for operating a sideline business or selling in quantity to others.

Thus, a final inventory control measure to be added to any business—but especially to businesses that have a sizable number of expendables—is a daily sign-out log. Forget the "honor system," because that's just an invitation to steal. Instead, appoint a responsible person to make sure that each employee who needs an expendable item signs for that item.

Figure 6-3 presents guidelines for creating a daily "components" or "expendables" sign-out log. You can use continuation sheets that repeat the bottom part of the cover sheet shown. Make sure that the employee

Business Operating Supply and Equipment Use Log

Date and time ————————————————————————

Department ——————————————————————————

Management representative

————————————————————————————————

Accountability of items (general description)

————————————————————————————————

Accountability of items (general description)

————————————————————————————————

Printed name, department, time and date, signature

Items issued ——————————————————————

————————————————————————————————

————————————————————————————————

————————————————————————————————

Figure 6-3. Daily sign-out log.

responsible for the log staples the sheets together and turns them in each day, even if no supplies were issued or shown as issued. You can verify the log through the daily inventory reports and worksheets (prepared by employees other than the person responsible for issuing expendable items), using the guidelines shown in Figs. 6-1 and 6-2.

Collect the sign-out sheets daily and attach them to your other inventory control documents. It not only will serve as a fast way to spot "patterns" of theft or careless use of business assets but also as a sound deterrent to theft by eliminating employee anonymity. The log is a good management tool to control operating costs and often results in increased net profits.

Employee Pilferage Against Business Customers

There is one more important form of pilferage you need to consider, and that is employee theft from business customers. Although your customers will often sustain the direct loss, they will discover or suspect that your business is the culprit. Lawsuits may result, and future business may be lost. Word-of-mouth advertising best serves any business. However, it works two ways. News spreading about your business ripping off customers can send present and future customers fleeing to competitors.

Employee pilferage against customers typically occurs when an employee gains credit card numbers and then uses them in mail orders sent to a post office box or some other address that cannot be traced. Some employees divert the attention of a customer and then return another credit card (often counterfeit), retaining the one the customer gave them. The customer, confused or in a hurry, stuffs the card into a wallet or purse and leaves, without noticing the switch. Other employees get the customer to leave while the card is still in the imprint machine. The customer is unlikely to notice the missing card until trying to make a purchase at a later date. If the switch is discovered, the employee can "act stupid," and the customer will only be upset.

The employee who successfully pulls off this type of pilferage may sell the credit card on the black market or use it to make countless small purchases (to avoid notice) and then discard it after running up huge bills. Some employees use the carbons from a credit card transaction for mail order, and still others ask for a major card and record the number on a check, later copying it and using the information for mail-order fraud. This kind of pilferage happens extensively each year.

Other employee techniques against customers include overcharging, shortchanging, and even stealing from purses or shopping bags left on the counter while a customer goes into a dressing room with apparel. This type of pilferage is more widespread than is recognized and can create huge losses simply because customers identify your business as dishonest and shy away.

Security Technique 79: Institute Controls to Protect Your Customers

You can end employee pilferage against customers by implementing several strict policies. First, for a few dollars purchase several waste-basket shredders and place one next to each credit card imprint machine. Create a strict policy that all carbons from credit card trans-actions are to be placed immediately into the shredder in full view of the customer. Once information is shredded, access to card numbers is denied to employees or cleanup people.

Second, do not ask a customer for a major credit card to accept a check. It means nothing, because the person's credit card does not guarantee a check. Nor does a credit card protect you against forgers or bad check artists, who may have already acquired phony cards to overcome that obstacle. The only person this practice benefits is the employee pilferer or the cleanup person who uses the number fraudu-lently to create a liability and lost business.

Many businesses are being pressured into using electronic credit card machines. The cashier passes the card through a reader on the machine, the machine produces a printout of the sale, and the customer signs the printed single-copy slip (similar in width to an adding machine tape). Although many banks and credit card companies favor the system, believing it will reduce fraud, it can actually increase the opportunities. For one thing, even though the customer signs the machine-generated tape, it remains on file at the business. With the traditional imprinting of credit cards, the hard copy with the customer's signature must be deposited in a bank and it later returns to the customer in a monthly billing statement. Without the imprint of the credit card, the card num-ber has limited value to the pilferer.

Also, most electronic systems do not require the card to be "read" by the machine; instead, the cashier can type in the information on a key-board. Using information from the original slip, a pilfering employee can reenter the card number a day, a week, or even a month later and enter a purchase amount the same as a previous cash sale. The employee then places the slip in the cash register without ringing up the card and pockets the equivalent in cash.

Although the new electronic system is touted as a fraud stopper, the publicity seems mainly geared to pressuring business owners into accepting the $1000 price tag per machine. The machines can make credit card sales easier, but again *they do not stop fraud.* Also, by the time a customer complains about a billing, up to 2 months later—if the customer ever does complain—the employee who perpetrated the thefts will probably have moved on, or will just deny any involvement.

Hotels, restaurants, and other businesses that take deposits or orders by telephone face an added problem. An employee can take the credit card information by telephone and use it either on the conventional imprint machine or on an electronic machine. The employee can then (as in the retail store scam) repeat the transaction later for the same amount as a cash purchase, and pocket the cash.

You should also have a strict policy ensuring that employees do not allow customers to leave purses, shopping bags, or packages sitting around while they go elsewhere. If packages must be left for some reason, the employee should be required to summon another employee or even a manager immediately to witness that no one tampers with unattended personal items.

Another effective policy is to place a prominent sign on the front of the cash register telling customers that if they do not receive a receipt from the employee, the product or merchandise is theirs without charge. Follow up with a stipulation that the employee's mistake will be deducted from his or her salary. Any money that employees could pilfer from not ringing up a sale (resulting in no receipt) or from overcharging or shortchanging will clearly not reap them any profit; instead, it will cause employees to lose money and maybe their jobs.

The best preventive policies for your business will suggest themselves as you carefully assess your vulnerability. If you later discover new schemes stemming from the ingenuity of a pilfering employee, immediately establish equally creative countermeasures to eliminate the opportunity and especially to eliminate any personal gain for the employee. You will quickly eliminate the problem.

Quick Reference Guide to Chapter 6: Preventing Employee Pilferage

- Recognize the problem! Employee pilferage could move your business into the red when it would otherwise enjoy a sound net profit.

- Recognize the many techniques employees use to pilfer your business assets and turn them to your advantage by creating effective countermeasures.

- Conduct regular pilferage vulnerability assessments in your business operations.

- Create an effective inventory system that complements your daily interest in business income and that immediately alerts you to an existing or potential pilferage problem.

- Include your employees in the inventory control program. Participation increases their awareness and responsibility. It also supplies you with a verification record.

- Establish strict and effective policies and countermeasures that protect your business assets. Extend your controls to pilferage of customers, which could create huge losses from lost business and lawsuits.

7
Eliminating Bad Checks and Credit Card Fraud

Each year bad checks and credit card fraud account for business losses that stagger the imagination. Specific figures cannot be calculated exactly, because many businesses do not bother to report these crimes to police or to credit card companies. Often, bad checks and returned credit card vouchers wind up in a drawer; and these losses are not deducted from net profits.

From the incidents that are reported, we do know that losses from accepting bogus credit cards top $2 billion each year, while bad checks cost businesses at least $5 billion. Various government agency programs have had little success in thwarting this type of fraud. You can do something about it. Take preventive action by eliminating the opportunity.

The Bad Check Artist: A Profile

People who deliberately pass bad checks may make all or part of their living from this crime. In a given year, about 34 billion checks move through businesses. To spot the real ones from the phonies, you need to know what to look for when accepting checks.

Although many businesses believe they have a sound prevention plan, they actually supply the bad check artist with extraordinary opportunities. For example, I have written checks in a variety of stores across the country. Often, when I present my driver's license for identification, the clerk rejects it and asks for one or two credit cards. If I were searching out "marks" for a bad check scheme, I would clearly choose this type of business as one of my victims. A professional check passer will have counterfeit but legitimate-appearing identification and will give a convincing reason why you should accept the check. A bogus credit card is easier to create than a bogus driver's license.

Professional bad check passers research the targeted community and businesses by making legitimate transactions. For example, the check artists will come into town and attempt to establish their credentials as honest and forthright citizens. They make the rounds of targeted establishments—usually smaller, community-oriented businesses in affluent areas—and write small "good" checks. They also establish a personal rapport with business owners, managers, and clerks. Next they skillfully become a part of the community, often under the guise of preparing to open a business themselves.

Although such a scheme might sound like it takes a long time, this is not the case, especially in suburban or rural areas. Once the con artists have established credibility with their skills of persuasion, they strike. They carefully choose the timing of their purchases and select merchandise that they can consume personally or easily sell elsewhere. Here is a typical scenario.

Mr. Smith arrived in an affluent town on a Monday morning and left 3 weeks later after successfully passing $180,000 worth of bad checks. Upon arrival, driving a conservative new car (later discovered as rented from a nearby airport for cash under a fictitious name), Smith checked into a local inn and paid for the room. He began making the rounds of local businesses. In the beginning, Smith said little but made small purchases (a package of gum, a necktie, some writing paper) for cash.

The following day, Smith called on the local chamber of commerce and a couple of banks to discuss transferring a large sum of money. Each bank he visited, anticipating a large wire transfer deposit, opened a business account for him with a small deposit. Now Smith had two legitimate checking accounts in a business name. His next act involved going to the nearby city and having checks printed, to be picked up 2 days later. Back in the small town, he made the rounds of realtors, supposedly seeking to buy both a residential property and a commercial property. Well-dressed and convincingly affluent, but clearly projecting an image of a conservative businessperson, Smith soon had the grapevine buzzing that a new entrepreneur was in town.

For the next 2 weeks Smith continued contacting realtors, visited his targeted stores, and skillfully discussed his planned business opening. He selected a venture that had appeal but would not compete with any other business in town. During this time, Smith began making small purchases with his "local checks," which promptly cleared. He also ensured that owners, managers, and clerks clearly saw his many credit cards, although he never used them.

Finally, the time came that Smith had waited for: a 3-day holiday weekend. He told the innkeeper that he had leased a house but instead moved to a motel several miles away. After checking out, Smith began to make the rounds of local businesses, starting about noon on Saturday and continuing through the following Monday evening (the holiday). He bought costly watches and a variety of other expensive items from his target stores. He talked to the clerks or managers he had come to know well and used a pretext for the purchase, claiming that he had to send a gift to someone important or to a family member. Because it was a holiday weekend and banks were closed, Smith asked for a moderate amount of cash above the purchase and easily received it.

On Monday evening, Smith drove to the airport, turned in the rental car, and took a flight to a city 2500 miles away. On Tuesday, he contacted a fence that he knew and sold off the items purchased with the bogus checks. He then took another flight to a city 1500 miles away, where he had a modest apartment. There he relaxed and destroyed the leftover checks, receipts, and other items that might link him to the town he ripped off. Smith then counted his net profits of just over $100,000—not bad for 3 weeks' work and modest expenses.

After some rest, Smith repeated his actions in a new town using a different name, new ID cards, and a fresh scheme. For several years he operated in this manner across the country, making a total of about $1.5 million! He would have continued to prosper had he not had a car accident on the way to the airport during one of his trips. Broken bones and a minor head injury left him hospitalized. He remained unconscious for a few days and recovered in about 3 weeks. Ironically, the hospital was in the same town where Smith had just passed a series of bad checks.

After the accident, police towed in Smith's wrecked rental car and inventoried his personal effects. Later, when merchants filed complaints of bad checks, police detectives knew Smith was the culprit. Fingerprints linked him to a series of other crimes, and police information networks soon established Smith as wanted across the country.

Security Technique 80: Check It Out

Since checks remain the most popular form of payment for goods and services in the country, your business probably depends on accepting them. The following countermeasures will help you avoid becoming the target of a professional bogus check passer like Smith. If you can implement these policies in your business, you can go far to prevent the bad check passer from even trying to make you a victim.

Begin by posting a sign on or alongside your cash registers showing clearly what you need if a patron wants to give you a check. Ensure that your managers and employees allow no exceptions. Inform them that any violation of your policy makes them "personally" liable for the check. Should they "feel sorry" for a customer who, like Smith, has a good story or simply fail to "check it out" properly, they will be responsible. The signs you post should warn the check passer of these policies.

Most supermarkets today require a courtesy check-cashing card for patrons. You should also create this policy. Note on your sign that checks will be welcomed if they are accompanied by a check privilege card issued by your store. Customers wanting to pay by check must fill out a form and pick up their check privilege card 3 business days later. This gives you time to verify the information they supply and helps to weed out bad check passers or those who might otherwise qualify but "kite," or those who regularly draw checks against insufficient funds. If someone bounces a check after receiving your check card, you and the police will have sound information and thus have a strong chance of recovering your money. Obviously, someone like Smith would not be eligible for your check-cashing card and, recognizing this barrier, would not target you. However, remember that you want to stop all those who pass bad checks. The more "fair" but "rigid" rules you create, the less likely you are to become a victim.

If a person wants to write a check for "immediate" goods and services (to walk out of your business with), you should require that the funds be drawn on a local bank. If customers pay by check for items they do not take with them, do not deliver the goods until the check clears.

Ask your check-writing customers to present a valid driver's license from your state—one that shows a "local" street address, not a post office box! This will thwart nearly all professional bad check artists. It is difficult to forge a driver's license with a local address, because the location can be easily checked out. Few con artists will go to the trouble it entails. Also, to obtain a local driver's license, they will have to surrender any other state license. If customers claim not to have a license, ask for other identification, and even for written tests and pictures. All these measures threaten the anonymity that is so important to the con artist's success.

Ask check-writing customers for a local, "verifiable" telephone number. Here you create still another barrier to the check passer. Unless the person is a familiar face in the community, or known to be a resident, verify the number in the telephone book or call information to ask for the telephone number of "Mr. or Ms. Smith." If it is unpublished, the operator will state that "Mr. or Ms. Smith" has a telephone but it is not listed. At least you will know with reasonable certainty, providing the other requirements are met, that the person has a legitimate standing in the community.

When the check is a "business check," follow the same process. Also, ask the bank to verify how long the account has been open and whether the check writer has authority to issue a check drawn on the business. If the purchase is made during nonbanking hours and you are not familiar with the business in question, do not accept the check. For example, your policy could read: "No business checks from other than known, established local businesses will be accepted without verification from the bank."

Security Technique 81: Circumvent Forged Checks

Even with rigid controls, your business might fall victim to forgery from customers who meet all your requirements—or from professionals who decide to go to the trouble of meeting local requirements, perhaps to make one large "score" before retiring to a tropical island. Because a scant 2 percent of check forgers are ever identified and convicted, the odds of success clearly favor the crook. When they are convicted, forgers rarely serve more than 5 years. Considering the money to be made by an industrious forger who aggressively plies his or her trade, the profits far exceed the risk.

Check forgery used to require money and sophistication. However, with today's color copiers, photo offset machines, and desktop publishing equipment, forgery is easier than ever. The forger need only "scan" a check into a computer, digitize it (using a variety of enabling software programs), and reprint it on a quality laser printer.

The most common type of forged check is one on which the person changes the printed information. Figure 7-1 shows a typical check. It is not difficult to spot a forgery using the following techniques.

Check for Perforations

Nearly all checks, except government Treasury checks, have at least one perforated edge from where the check was pulled from a checkbook, payment stub, or roll of continuous-feed computer paper. Even a cashier's check issued by a bank will have a perforation where it was separated from the customer's and bank's copies. Forged checks cre-

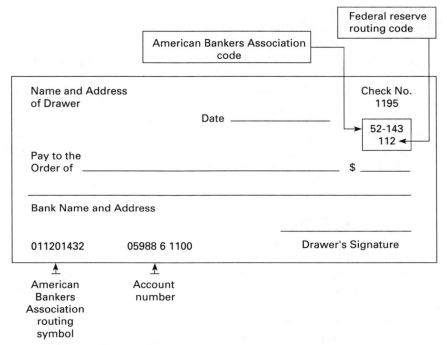

Figure 7-1. A typical check.

ated by a copier, offset machine, or computer will rarely have a perforated edge. Nor will sophisticated forgers go to the trouble of creating perforations. Why bother when methods such as persuading the targeted business to take the check are easier?

Verify the Federal Reserve District Number

A nine-digit American Bankers Association routing symbol appears on the bottom left corner of every check. The first two digits designate the Federal Reserve district where the bank is located and where the check will eventually be processed. There are 12 districts in the United States, numbered 01 to 12 from the east to west coasts. Thus, a legitimate "local" check in California might have the first two digits of 12. Suppose a forger altered it to read 02. The person accepting the check may not know the districts or notice that the routing code reads 02, not 12. The cashier notes only that the check is drawn on a local bank. Changing the routing number to 02 offers the forger an advantage because the check will, upon deposit, be routed to Federal Reserve District 02, which is Manhattan, New York. That gives the forger some 5 to 10 days to go throughout California cashing checks without worrying that any of them will bounce and alert authorities to forgery.

So when you accept a check, even if it looks authentic and the person handing it to you has met all your requirements, be sure to look at the Federal Reserve code. Find out which Federal Reserve district you're in. If a check from inside the district has a different number, it is a fake. When a check is drawn on a savings and loan account, the Federal Reserve will add 20 to the routing number. The routing number on a check from a California savings and loan, for example, would start with 32 instead of 12.

Examine Other Codes

Other code numbers on a check supply valuable information that can help you spot forgeries. Figures 7-2 through 7-4 provide details. A quick scan of these coding systems could save you a sizable chunk of net profits.

The check acceptance card system discussed earlier will also make you less vulnerable to check forgers. Stay alert, however. Some people take the trouble to meet all the requirements only to slip in that one big forgery before skipping town. Also, most forgers prefer to create illusions, not paper trails, and develop considerable skill in remaining nondescript so that victims have difficulty remembering them. A forged check will take days and sometimes weeks to return to a business. It is rare that the business owner, manager, or clerk can remember what the forger looked like more than a day or two after accepting a bad check.

Security Technique 82: Recognize Phony Business Checks

Phony business checks pose a threat not only to retail businesses. Wholesalers and other suppliers that usually receive payment from business customers by check are also vulnerable. Typically, the scam artists create a "paper" company designed not to conduct legitimate business but to steal the assets of others. For example, a person claiming to be vice president of ABC Corporation enters your wholesale supply business and announces that ABC has just come to town and selected you as its supplier. The VP impresses you immediately by dropping a familiar business name or two and making it clear that ABC is a conservative "pay as we go" operation and always pays COD.

(text is continued on page 148)

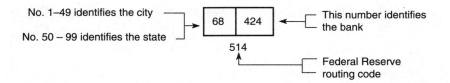

No. 1–49 identifies the city ⟶ 68 | 424 ⟵ This number identifies the bank

No. 50 – 99 identifies the state

514

Federal Reserve routing code

The number 68–424 is identifiable as follows:

68 – State of Virginia

424 – Arlington Trust Company, Arlington, Virginia

(a)

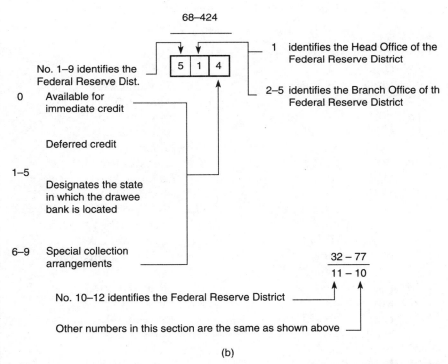

68–424

No. 1–9 identifies the Federal Reserve Dist.

5 | 1 | 4

1   identifies the Head Office of the Federal Reserve District

2–5   identifies the Branch Office of th Federal Reserve District

0   Available for immediate credit

Deferred credit

1–5

Designates the state in which the drawee bank is located

6–9   Special collection arrangements

32 – 77
11 – 10

No. 10–12 identifies the Federal Reserve District

Other numbers in this section are the same as shown above

(b)

Figure 7-2. (*a*) American Bankers Association code. (*b*) Federal Reserve routing code. The number 514 is identified as follows: 5=fifth Federal Reserve District; 1=head office in Richmond, Virginia; 4=deferred credit and the State of Virginia.

Figure 7-3. American Bankers Association index to prefix numbers of cities and states.

Prefix Number Index

Numbers 1 to 49 are prefixes for cities

Numbers 50 to 99 are prefixes for states

Prefix numbers 50 to 58 are eastern states

Prefix number 59 covers Alaska, American Samoa, Guam, Hawaii, Puerto Rico, and Virgin Islands

Prefix numbers 60 to 69 are southeastern states

Prefix numbers 70 to 79 are central states

Prefix numbers 80 to 88 are southwestern states

Prefix numbers 90 to 99 are western states

Prefix Numbers of Cities in Numerical Order

1	New York, NY	26	Memphis, TN
2	Chicago, IL	27	Omaha, NE
3	Philadelphia, PA	28	Spokane, WA
4	St. Louis, MO	29	Albany, NY
5	Boston, MA	30	San Antonio, TX
6	Cleveland, OH	31	Salt Lake City, UT
7	Baltimore, MD	32	Dallas, TX
8	Pittsburgh, PA	33	Des Moines, IA
9	Detroit, MI	34	Tacoma, WA
10	Buffalo, NY	35	Houston, TX
11	San Francisco, CA	36	St. Joseph, MO
12	Milwaukee, WI	37	Fort Worth, TX
13	Cincinnati, OH	38	Savannah, GA
14	New Orleans, LA	39	Oklahoma City, OK
15	Washington, DC	40	Wichita, KS
16	Los Angeles, CA	41	Sioux City, IA
17	Minneapolis, MN	42	Pueblo, CO
18	Kansas City, MO	43	Lincoln, NE
19	Seattle, WA	44	Topeka, KS
20	Indianapolis, IN	45	Dubuque, IA
21	Louisville, KY	46	Galveston, TX
22	St. Paul, MN	47	Cedar Rapids, IA
23	Denver, CO	48	Waco, TX
24	Portland, OR	49	Muskogee, OK
25	Columbus, OH		

Figure 7-3. (*Continued*) American Bankers Association index to prefix numbers of cities and states.

Prefix Numbers of States in Numerical Order			
50	New York	74	Michigan
51	Connecticut	75	Minnesota
52	Maine	76	Nebraska
53	Massachusetts	77	North Dakota
54	New Hampshire	78	South Dakota
55	New Jersey	79	Wisconsin
56	Ohio	80	Missouri
57	Rhode Island	81	Arkansas
58	Vermont	82	Colorado
59	Alaska, American Samoa, Guam, Hawaii, Puerto Rico, and Virgin Islands	83	Kansas
		84	Louisiana
		85	Mississippi
60	Pennsylvania	86	Oklahoma
61	Alabama	87	Tennessee
62	Delaware	88	Texas
63	Florida	89	District of Columbia
64	Georgia	90	California
65	Maryland	91	Arizona
66	North Carolina	92	Idaho
67	South Carolina	93	Montana
68	Virginia	94	Nevada
69	West Virginia	95	New Mexico
70	Illinois	96	Oregon
71	Indiana	97	Utah
72	Iowa	98	Washington
73	Kentucky	99	Wyoming

Figure 7-4. Federal Reserve banks and branches.

1	Federal Reserve Bank of Boston Head Office	$\dfrac{5\text{-}1}{100}$
2	Federal Reserve Bank of New York Head Office	$\dfrac{1\text{-}120}{210}$
3	Federal Reserve Bank of Philadelphia Head Office	$\dfrac{3\text{-}4}{310}$

(*Continued*)

Figure 7-4. (*Continued*) Federal Reserve banks and branches.

4	Federal Reserve Bank of Cleveland Head Office	$\dfrac{0\text{-}1}{410}$
	Cincinnati branch	$\dfrac{13\text{-}43}{420}$
	Pittsburgh branch	$\dfrac{8\text{-}30}{430}$
5	Federal Reserve Bank of Richmond Head Office	$\dfrac{68\text{-}3}{510}$
	Baltimore branch	$\dfrac{7\text{-}27}{520}$
	Charlotte branch	$\dfrac{66\text{-}20}{530}$
6	Federal Reserve Bank of Atlanta Head Office	$\dfrac{64\text{-}14}{610}$
	Birmingham branch	$\dfrac{61\text{-}19}{620}$
	Jacksonville branch	$\dfrac{63\text{-}19}{630}$
	Nashville branch	$\dfrac{87\text{-}10}{640}$
	New Orleans branch	$\dfrac{14\text{-}21}{650}$
7	Federal Reserve Bank of Chicago Head Office	$\dfrac{2\text{-}30}{710}$
	Detroit branch	$\dfrac{9\text{-}29}{720}$
8	Federal Reserve Bank of St. Louis Head Office	$\dfrac{4\text{-}4}{810}$
	Little Rock branch	$\dfrac{81\text{-}13}{820}$
	Louisville branch	$\dfrac{21\text{-}59}{830}$
	Memphis branch	$\dfrac{26\text{-}3}{840}$

Figure 7-4. (*Continued*) Federal Reserve banks and branches.

9	Federal Reserve Bank of Minneapolis Head Office	$\dfrac{17\text{-}8}{910}$
	Helena branch	$\dfrac{93\text{-}26}{920}$
10	Federal Reserve Bank of Kansas City Head Office	$\dfrac{18\text{-}4}{1010}$
	Denver branch	$\dfrac{23\text{-}19}{1020}$
	Oklahoma City branch	$\dfrac{39\text{-}24}{1030}$
	Omaha branch	$\dfrac{27\text{-}12}{1040}$
11	Federal Reserve Bank of Dallas Head Office	$\dfrac{32\text{-}3}{1110}$
	El Paso branch	$\dfrac{88\text{-}1}{1120}$
	Houston branch	$\dfrac{35\text{-}4}{1130}$
	San Antonio branch	$\dfrac{30\text{-}72}{1140}$
12	Federal Reserve Bank of San Francisco Head Office	$\dfrac{11\text{-}37}{1210}$
	Los Angeles branch	$\dfrac{16\text{-}16}{1220}$
	Portland branch	$\dfrac{24\text{-}1}{1230}$
	Salt Lake City branch	$\dfrac{31\text{-}31}{1240}$
	Seattle branch	$\dfrac{19\text{-}1}{1250}$

The phony executive then tells you that within a few days, after ABC gets its offices and other operations in place, a large order will come your way. He or she assures you that this order will be the first of many and that it should be billed COD. A few days later, the same VP calls to say a large order is being faxed and that the head office is putting on pressure to get the branch operational by the weekend. The phony VP suggests that if you put the order together and have it on the loading dock, one of ABC's trucks will pick it up. The driver will be carrying a company check and will fill in the amount of the invoice when the items are received. You can't lose, and there's no reason to go through the usual credit check on the company.

A week later, you receive the grim news that not only did the check issued by ABC Company bounce, but the bank never heard of the business. After you notify the police, a local detective adds to your worst fears: The phony VP has hit many businesses in the same way, and the address noted on the checks has been traced to a vacant warehouse. Beyond several large worthless checks, the police have little to go on. Police suspect that two or three people, armed with phony tax numbers, business cards, and checks, stole items from wholesalers and then left town to sell the merchandise in a distant city.

Variations of this scam involve having you fill an order on the spot. Typically, the scam artists give you a computer-printed or check-writing machine business check that makes them appear to represent a bona fide commercial operation. The check may even be countersigned (by two executives) for greater "credibility."

Remember that printed checks are available from scores of legitimate sources, including mail order. These sources have no obligation to verify the legitimacy of a business or its bank account number. Before accepting a business check, verify that the business does in fact exist. Check your state incorporation office for registration or incorporation. All it takes is a telephone call. Even an out-of-state company must register to do business legally in another state. Also, check with your local chamber of commerce, with city business-licensing agencies, and with the bank that the check is drawn on before allowing your assets to go anywhere.

Security Technique 83: Spot Bogus Traveler's and Cashier's Checks

Traveler's and bank-issued cashier's checks were once considered the bastions of reliability in the business world. With modern technology, however, they too have become tools of the forger.

Traveler's Checks

Traveler's checks offer con artists two advantages. First, a traveler's check is welcomed in most business operations. Second, few people are familiar enough with all their various formats to detect a counterfeit.

To create a genuine-looking traveler's check, the passer purchases some legitimate ones and uses them to generate the phonies. Usually, the defrauder comes into a business during a busy time and makes a significant purchase of one or several items—creating a sale that the business does not want to lose. The check passer offers to pay cash but suddenly discovers that he or she does not have enough. The defrauder then asks if it would be acceptable to pay with a traveler's check.

Skilled check forgers use a well-known issuer, such as American Express, and make the checks in a variety of denominations. In this kind of scam, the purchase will be over $100, probably $120, and the denominations of the traveler's checks will be $100 each. The defrauder will also have a forged driver's license and forged credit cards, in the same name as that used on the traveler's checks. The person will sign two of the $100 traveler's checks and receive the merchandise, chosen carefully for personal use or for selling after the scam, along with $80 in cash. This type of check forger works relentlessly during weekends and other times when banks are closed and uses that as an excuse for giving you large-denomination traveler's checks. By the time you learn that the checks are bogus, the scammer is enjoying the sea breeze in Bermuda.

What can you do to protect your business against this type of fraud? Begin by buying a supply of the commonly used traveler's checks through your local bank, and keep one of each check at each cash register in your business. Do not secretly examine the traveler's checks handed to you. Remember that prevention—eliminating the opportunity—will discourage all but the most brazen of defrauders. Instead, keep the sample checks in a folder, and when a person offers to pay by traveler's check, bring out the folder and announce that you need to see if the check is one that your business will accept. In bold lettering in plain sight within the folder, instruct cashiers to validate traveler's checks by comparing them with those in the folder and by checking them against an attached list. Add a final note that instructs cashiers to "call police should a forgery be suspected."

Remember to keep your instructions in bold lettering so the person handing over the check can easily see them. The folder serves both as a deterrent and as a way for cashiers to quickly validate the traveler's check. Normally, check passers will "change their mind" and suddenly find enough cash to pay for the purchase, or just leave your business in a hurry. Don't let the matter rest there. Notify the police of your suspicions. You protect your assets this way, and you may also aid other businesses. Your efforts can help the police get a forger off the streets.

Another technique for thwarting losses from bogus traveler's checks is to be sure that the person signs the check in front of you. If the transaction happens during normal banking hours and you have the slightest doubt, announce that you do not have enough cash reserves to accept the traveler's check above the purchase amount. Suggest that the customer cash the checks at a nearby bank, and offer to hold the purchase until he or she returns. Check forgers survive not only on using or selling the product they purchase but also on receiving cash. When setting up your preventive measures, picture yourself as the traveler's check forger. Consider what would make your business a good, risky, or impossible target, and then establish countermeasures that will protect your business assets.

Cashier's Checks

Cashier's checks can easily be forged, and my best advice is that if you don't know the person offering a large-denomination cashier's check, contact the issuing bank. If you cannot do that, do not accept it. It is better to lose a sale than "give away" your business assets. During normal banking hours, suggest that the patron cash a cashier's check at a bank. If the patron voices "indignation" about your skeptical attitude or refusal to accept the check, you can be reasonably sure that he or she is trying to steal your business assets. Legitimate customers might show some frustration or dissatisfaction at the inconvenience; however, most will understand your reluctance. By contrast, check passers will stage a dramatic outburst to conceal their scam, and it frequently intimidates the cashier enough to take a chance just to appease them. Do not fall for that type of intimidation.

The Scope of Credit Card Fraud

Credit card fraud stems from four major sources: stolen cards, counterfeit cards, cards that have exceeded their limits, and cards used after cancellation. Many businesses verify the status of cards through a computerized network. Each card must pass through a "reader" that tells the cashier if the card has exceeded its credit limit or has been reported stolen or lost.

For the skillful defrauder, verification machines become a friend instead of a threat. One of the major weaknesses in the reader system is a lack of personal identification numbers (PINs), such as those used for bank automatic teller machines (ATMs). Although it's possible for a person to obtain the PIN for a credit card that is stolen or lost, it's not probable. Currently, however, credit card machines do not have a system for verifying the identity of the person handing over the card. If a card has been lost or stolen (but not yet reported as such), the machine will give the user a "green light." If your business uses the electronic system and you receive a green light, make sure that you will not be debited by the bank or credit card company if the card is not valid for a purchase.

Most businesses do not use electronic readers, either because the service is too costly or because it is not available in their area. Independent or small firms, in particular, continue to use the traditional imprint machines, and thus become primary targets for credit card fraud.

If you use the standard imprint machines, you will receive a bulletin periodically showing numbers of credit cards you cannot honor. However, ensure that your bank or the credit card company will guarantee the purchase even though it does not exceed the amount necessary to call in for authorization. Also, if you have any doubts about the validity of a card, call for authorization regardless of the amount of purchase.

Security Technique 84: Recognize Common Credit Card Scams

With the proliferation of credit cards from financial institutions and even big businesses, scams and counterfeiting have become lucrative tools of criminals. Since most people now carry several credit cards, theft has become popular with many categories of criminals. More hardened types prefer to counterfeit their cards. A counterfeit card removes limitations and reduces the risk of the criminal being caught with a stolen or lost card. Counterfeiters create unlimited opportunities, and after vigorously working one area, they move on to another and may eventually return to those businesses that were targeted several years earlier.

Counterfeiters work from existing cards. They use carbons or electronic card machine tapes from legitimate transactions. They imprint the counterfeit card with the names and numbers of active accounts in good standing. Many counterfeiters use the cards personally or operate within an organized group. Other counterfeiters sell the cards to organized groups along with bogus driver's licenses or other identification documents.

Legitimate cards issued to applicants may be "waylaid" by credit card thieves. They may steal from legitimate manufacturing areas, from the mails, or from mailboxes after delivery before the applicant arrives home. The credit card defrauder can use stolen cards safely for about a month before notices go out to merchants.

The perpetrators of credit card fraud do not fit any mold. They may be devoted elderly couples, young people impersonating busy executives, or homemakers picking up extra money. Although businesses confront a variety of credit card scams, counterfeiting creates the largest asset loss, since no protection is offered by the issuing institution. When a card is lost or stolen and your business has not yet received a cancellation notice (i.e., the printed booklet you receive regularly), you are usually not liable.

Security Technique 85: Reduce Credit Card Fraud

Your first step in creating effective countermeasures at your place of business is to contact the various issuing agencies for assistance. Most credit card companies have a variety of safeguards that will help you and your employees detect counterfeit credit cards. For example, credit card logos often have holographic (three-dimensional) symbols that can be seen only under an ultraviolet light. Other fine details, like those found in genuine currency, also help you detect counterfeit cards. You should install an ultraviolet light at each cashier position for detecting counterfeit currency as well as counterfeit credit cards. The ultraviolet lights come in small, portable configurations that are easy to install and use. They are also an attractive addition to your checkout station.

If you are still using imprint machines, consider becoming linked to electronic credit card identification systems. There are some advantages. These systems operate from the encoded magnetic (black) strip on the back side of a credit card. The electronic machine identifies the codes by reading the magnetic strip. Counterfeit cards cannot duplicate these codes effectively enough to pass through the readers. Also, cards that have been reported stolen or lost will immediately become part of the electronic database. These systems supply the approval code number that will protect your business from loss even if the card does not appear in the database as discontinued.

Quick Reference Guide to Chapter 7: Paper and Plastic Fraud

Here are some tips to avoid falling prey to the bad check artist:

- Issue a check privilege card at your store.
- Request a driver's license showing a local street address (not a post office box) as identification.

- Request a local, *verifiable* telephone number.
- Verify business names and locations.
- Examine the quality of paper the check is written on and look for perforations along one side.
- Verify the Federal Reserve district number and other coding systems on the check.

The following signposts will alert you to possible credit card fraud and help you avoid becoming a business victim:

- The credit card has expired or is not yet valid.
- Alteration of the card is obvious.
- The card is not signed, or the signature differs significantly from the one on the transaction slip.
- The customer rushes in during busy hours, selects purchases rapidly, and keeps the amount just under the floor limit.
- The customer's attire and appearance are inconsistent with purchases or with the type of card presented (e.g., a shabbily dressed person presenting a "gold card").
- The cardholder asks you to split the purchase between charge slips, maybe in an attempt to forestall an authorization call to the issuer.
- The cardholder tries to rush or intimidate the cashier.
- The cardholder makes purchases, leaves the store, and returns for added purchases.
- The cardholder makes multiple purchases, all under the floor limit.
- The cardholder purchases many of the same items but in different colors and sizes.

8
Spotting Counterfeit Cash and Bogus Discount Coupons

According to the U.S. Secret Service, counterfeiters create about $245 million of bogus U.S. currency each decade. Over $31 million is successfully passed to individual victims, merchants, and banks. In past years, currency counterfeiters relied on "plates" that needed meticulous engraving by skilled craftspeople. Modern technology has created a way for nearly any person so inclined to pass off counterfeit currency to unsuspecting businesses that become alert to the problem only when they are notified by the bank. Under federal law, anyone possessing counterfeit currency and coins must notify the authorities and relinquish the money to the U.S. Secret Service.

The Mechanics of Counterfeit Currency

It is easier to combat business theft from counterfeiting when you understand the methods used by currency counterfeiters and become aware of the government safeguards that help you spot counterfeit money. The following guidelines will help you apply the recognition techniques.

Paper

The paper for U.S. currency is manufactured to secret specifications by a special process, under the supervision and protection of the government. During the pulp stage, distinctive red and blue fibers are engrained into the paper. The fibers are distributed at random throughout the paper by the mixing process. Some fibers are on the inside, some are on the surface, and others protrude through the surface.

Design

Various parts of the currency design are made on separate plates by different crafts workers, using varied methods. The separate plates are then assembled on a master plate with a transfer press. The image on the master plate is transferred to a steel roll that is used to produce a master, 32-subject, multiprinting plate. This plate is used for printing bills. The procedure ensures uniformity of design no matter how many plates and presses are used.

Engraving

It is impossible to engrave a perfect duplicate of a genuine bill. Several parts of a genuine bill are hand-engraved by masters. The hand-engraved portions are then combined with other parts of the design that have been produced by machinery. Thus, the prospective counterfeiter is faced with having to reproduce the work of not one but several engravers, each with a distinctive style, as well as duplicating the work of precision machinery. The counterfeiter must also cope with the problem of reproducing, with exactness, the proper length and shading of lines, the expressions on portraits, and the scrollwork—in short, forgery must be perfection.

Ruling Machine

The ruling machine produces the lines composing the background in the oval of the portrait on a bill. These lines are of uniform thickness from one end to the other, and the space between the lines does not vary. Except for the areas where contrast is desired between the background and the dark tones of the portraits, the depth of the lines remains uniform. The ruling machine is used for other designs on bills, including the background lines in the scroll beneath the portrait containing the denomination of the bill in letters.

Geometric Lathe

The majority of fine lacy designs on the borders of bills are created by a geometric lathe. Visual examination of a genuine bill will reveal that the lathe work consists of white lines, with the ink appearing in background areas.

Printing Process

Two of the basic processes used in printing U.S. currency are the intaglio process and the typographic process.

Intaglio.   The basic face and the entire reverse of all genuine currency are now printed by intaglio. In this process, the design is engraved into the surface of the plate. The ink is thicker and coarser than planographic (offset) printing ink, which is commonly used by counterfeiters. Under a stereoscopic microscope, the ink lines on a genuine bill appear to be mounds of ink piled on top of the paper, and the grains of the coarse ink become visible.

In the intaglio process, the ink fills the engraved depressions. When the paper is forced against the plate with considerable pressure, the resulting design is composed of the characteristics of the mounds of ink deposited by the plate. The raised printing, at least on a new bill, resembles embossed printing on invitations, letterhead stationery, and greeting cards.

Typography.   The remaining features on the face of bills (those that are changeable) are printed typographically. Typographic printing is flat rather than raised and can be easily identified under magnification. The printed lines of the typeface are pressed into the paper, and the pressure causes a heavier deposit of ink around the border or outside of the letter or design.

Types of Paper Currency

There are three basic types of paper currency in circulation: silver certificates, U.S. notes, and Federal Reserve notes. Silver certificates are no longer printed or issued, and are currently being withdrawn. Most remain in the hands of collectors.

Security Technique 86: Identify the Characteristics of U.S. Paper Currency

Because Federal Reserve notes are the most common paper currency, they also are the most commonly counterfeited. Here are several sight recognition methods.

- On the left of the currency portrait, you will find the seal of the Federal Reserve Bank and in the middle of the seal is a letter or number identifying the bank of issue. Most notes today bear a letter (the use of numbers was discontinued in 1935).

- Federal Reserve notes have a number in each of the four corners of the white area on the face. These numbers range from 1 to 12 and, like the letter in the Federal Reserve seal, correspond to the 12 Federal Reserve districts.

- The prefix letter of the serial number on Federal Reserve notes is the same as the district letter.

- The suffix letter of the serial number changes with every 100,000 notes created.

Security Technique 87: Examine the Federal Reserve Bank Identifier, Letters, and Number Codes

Table 8-1 will help you identify the common Federal Reserve Bank letter and number codes that are present on genuine U.S. currency. The bank letter appears in the center of the Federal Reserve seal and also as the prefix letter in the serial number. The bank number appears in the four inner corners on the face of the note.

Table 8-1. Federal Reserve Codes

Bank	Bank letter	Bank number
Boston	A	1
New York	B	2
Philadelphia	C	3
Cleveland	D	4
Richmond	E	5
Atlanta	F	6
Chicago	G	7
St. Louis	H	8
Minneapolis	I	9
Kansas City	J	10
Dallas	K	11
San Francisco	L	12

Security Technique 88: Recognize the Color of the Treasury Seal and Serial Number

You can often identify counterfeit currency from the color of Treasury seals and serial numbers. Table 8-2 can help you recognize the differences.

Security Technique 89: Learn Currency Portraits and Back Designs

Each denomination, regardless of class, has a prescribed portrait and back design selected by the Secretary of the Treasury. Table 8-3 lists those you will commonly encounter in your place of business.

Security Technique 90: Examine Currency for Variable Features

Certain features of genuine currency vary from one series year to another. Many of the changes are instituted by the Secretary of the Treasury or by the Treasurer of the United States. The following vari-

Table 8-2. Seals and Serial Numbers

Class	Color	Denomination
Federal Reserve notes	Green	$1, $2, $5, $10, $20, $50, $100
U.S. notes	Red	$2, $5, $100
Silver certificates	Blue	$1, $5, $10

Table 8-3. Portraits and Back Designs

Denomination	Portrait	Back design
$ 1	George Washington	Great Seal of the United States
$ 2	Thomas Jefferson	Signing of Declaration of Independence
$ 5	Abraham Lincoln	Lincoln Memorial
$ 10	Alexander Hamilton	U.S. Treasury Building
$ 20	Andrew Jackson	White House
$ 50	Ulysses S. Grant	U.S. Capitol
$100	Benjamin Franklin	Independence Hall

ables will help you recognize the differences between counterfeit and genuine currency.

Serial Numbers and Series

As noted above, the prefix letter of the serial number on Federal Reserve notes always agrees with the letter on the Federal Reserve seal. If a Federal Reserve note is accidentally mutilated during manufacture, it is replaced by a note having a special serial number with a star instead of the *suffix* number. Mutilated silver certificates and U.S. notes are replaced by special notes having a star instead of the *prefix* letter.

The series legend indicates the year in which the design of a note was approved, *not* the year in which the note was printed. The suffix letter following the series year indicates a minor change in the design, such as a change in the signatures of the Secretary of the Treasury or the Treasurer of the United States.

Faceplate Numbers, Check Letters, and Backplate Numbers

The U.S. Secret Service keeps a record of the denomination, type of note, faceplate number, check letter, and backplate number of all known counterfeit issues.

The faceplate number is found in the bottom right corner of a genuine note and identifies the multisubject plate used to print the basic face design. The number is preceded by a small letter known as the check letter, which is coded to the particular image on the multisubject plate.

The backplate number is found in the bottom right inner area on the reverse of each genuine note and identifies the multisubject plate used to print the reverse of the note.

How Criminals Counterfeit Currency

Originally, the only method of counterfeiting paper money was from engraving heavy plates for printing presses. Most counterfeiters today use a photomechanical method known as offset lithography. The availability of prepared chemicals, low-cost photographic equipment, inexpensive photomechanical printing plates, and books on photomechanical printing processes makes it possible for unskilled criminals to produce good counterfeit notes in large quantities. A variety of other methods are used as well, as detailed below.

Counterfeit bills range from deceptive (hard to detect) to poor (easy to detect). A deceptive counterfeit note will usually take more than a cursory glance to distinguish it from a genuine bill. A poor counterfeit note is readily detected by anyone having a basic knowledge of genuine currency.

Offset Lithography

Presensitized aluminum plates for offset lithography were introduced about 1950. Ready-to-use preparations, with instructions on the containers, enable even a novice to develop the photographic image for a bill on a plate.

The presensitized aluminum plate is exposed under a photographic negative of the note. Chemicals on the pretreated plate react, and an image of the note appears on the plate. Before printing, the plate is moistened with water, which chemically repels the ink from the non-printing surface. The plate is then inked, with the ink adhering only to the part of the plate that is to print the design. This plate is transferred to a rubber blanket roller that prints the paper. The small, high-speed offset press permits rapid reproduction.

Copying Methods

Photography. Fairly deceptive counterfeit notes can be manufactured by straight photographic processes. In some cases, very thin photographic papers are used to make separate prints of the front and back, then glued together to complete the note. Or photographic papers sensitized on both sides are used to print the note on the same sheet of paper.

Colors in the seal and numbers, and on the back of the note, are produced by hand application of photographic toners or coloring tints. Direct color photography is normally not practical, because of its high cost, slow rate of production, and undesirable surface gloss.

Copier Machine. Counterfeiters may also create notes using color copy machines and high-quality paper. Such bills are usually passed to change-making machines, which scan them to determine their validity. If the bill appears genuine to the scanner, it will make change.

Alteration Methods

A counterfeiter may alter genuine notes to change their value, such as "raising" a $1 bill to a $10 bill. The altered notes are used in confidence games or sold to collectors. Here are some common methods of alteration.

Pen and Ink. The criminal removes the denomination markings or other note features with abrasives. The altered design is carefully drawn in and blended with the surrounding genuine features.

Paster Notes. The counterfeiter tears one or two corners from several genuine notes of the same denomination, accumulating a total of four corners. The counterfeiter then thins down the corners on a genuine note of a smaller denomination and pastes in the torn corners, using pen and ink to blend in the edges.

Pieced Notes. The counterfeiter tears a different section from a number of genuine notes of the same type and denomination until he or she has enough corresponding sections to piece together an additional note. The criminal takes care to remove no more than two-fifths of any of the genuine notes, so that they can be redeemed for full value.

Split Notes. The criminal splits the paper of genuine notes of several different denominations. He or she then pastes the front of the higher denomination to the back of the lower denomination, and vice versa. The criminal then attempts to pass both composite notes by showing the side with the higher denomination. In some cases, a counterfeit front or back is prepared to paste on the opposite side of the split genuine notes.

Bleached Notes. The counterfeiter bleaches a genuine note by removing all the ink until a piece of genuine currency paper is blank. He or she then uses a counterfeit plate and offset lithography to print a note of higher denomination on the paper. The obvious advantage of passing genuine paper makes this a popular counterfeiting method. In some cases, the criminal will bleach out only the denomination markings and portrait and then attempt to pass the altered bill to an unsuspecting victim.

Security Technique 91: Examine the Paper

You can quickly detect counterfeit currency by feeling the paper and comparing it with that used for genuine currency. Counterfeiters use a high-grade of bond paper, which they then immerse in fluid to create a worn look and a color resembling genuine paper currency. The counterfeit paper may have printed or hand-drawn red and blue fibers to simulate those of genuine bills. The fibers in a genuine bill can be lifted from the paper using a sharp instrument under a magnifying glass.

In genuine currency, the faceplate impression is larger than the backplate impression. Look for signs of watermarks on the paper by placing it under a strong light. Watermarked paper provides you with conclusive proof of a counterfeit bill. Also examine the edges of the paper for possible signs that two pieces have been glued together or that a genuine note was split by the counterfeiter.

Security Technique 92: Analyze the Ink

It is a myth that if the ink on a note rubs off on paper or cloth it is counterfeit. The ink on both genuine and counterfeit bills will rub off. The way to spot a bogus note is to check the brightness: Counterfeit ink is glossier than that used on genuine notes.

How Counterfeit Money Circulates

Much counterfeit currency (and some coins) travels without detection from person to person, through banking systems, and often across the country. Most counterfeiters focus their attention on large denominations of bills, such as $50 and $100. Others deal in large quantities of smaller-denomination bills, which receive much less scrutiny. These small bills usually pass through many people and businesses, making it impossible to trace them back to the counterfeiter or to the passer of bogus money.

A counterfeiter can move his or her work into circulation easily by using bogus $10 and $20 bills mixed with genuine currency. Typically, the counterfeiter purchases items that cost over $200, paying with a mix of genuine and counterfeit bills. Because cashiers and merchants usually suspect only larger denominations, the mix of genuine and counterfeit will be accepted easily. During the business day, the cashier might give out all or part of the counterfeit currency as change to customers. They in turn spend it at another store, and the cycle continues. Eventually, however, someone loses. A business will make a cash deposit and its bank will spot the counterfeit—most likely by using sophisticated scanning equipment.

Remember that if you detect a counterfeit bill handed over in payment for goods or services, it does not necessarily mean the customer has intentionally passed bogus money to you. Chances are that the person has no idea the money is *not* genuine. Counterfeiters or counterfeit-passing groups have developed skilled techniques and generally will not be caught in the act or identified later unless they stay in the same area and continue the scam.

Security Technique 93: Use a Counterfeit Bill Detector

Technology available to law enforcement and crime laboratories for decades is now in use in many businesses and provides a reliable and discreet means of detecting counterfeit bills. The ultraviolet (UV) light, built into a variety of configurations, can scan the bills quickly. If the currency is counterfeit, the ink and paper emit a telltale fluorescence. These UV lights can also identify bogus credit cards, traveler's checks, and altered or "write-over" signatures.

Figure 8-1 shows an ultraviolet currency scanner developed for commercial use. Counterfeit currency will glow when placed under a UV lamp. Genuine currency will not glow; the red and blue fibers embedded in the paper will show clearly. The various UV units in commercial use operate in a similar manner.

Figure 8-2 shows a discreet way of testing all currency accepted by a cashier. You should configure your scanning system so the customer cannot "hand" the money directly to the cashier. Instead, the customer should have to lay it on the counter. As your cashier slides the currency across the counter, it passes under the ultraviolet light. Should the bill be counterfeit, it will glow. In that event, the cashier should be trained to call for management to advise the patron.

It is important to create a business atmosphere that implies that you trust your patrons. You can disguise your UV system by using it as an advertising or instructional platform. For example, you might place a sign on it that shows your business check acceptance policy, or a placard announcing a sale in one of the store departments.

The Other Currency: Discount Coupons

Manufacturer's discount coupons have grown into big business and, unfortunately, criminals have found yet another way to defraud merchants. According to companies that act as clearinghouses for manu-

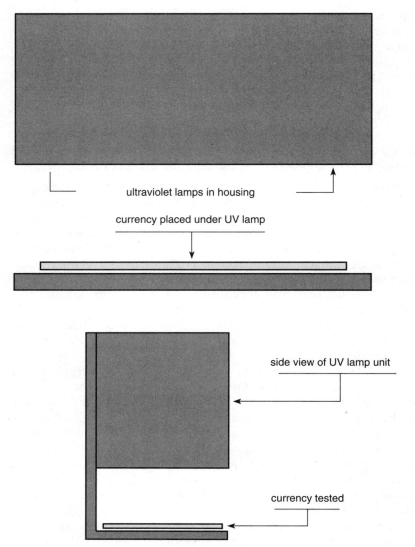

Figure 8-1. Ultraviolet currency scanner for commercial use.

facturer's coupons, about $800 million a year is paid to retailers and consumers who submit bogus or expired coupons. However, the manufacturers must necessarily crack down on coupon fraud, and when they do, your business could begin sustaining losses previously absorbed by the coupon issuer. When a merchant reduces merchandise and also expects to receive a small percentage from the issuing com-

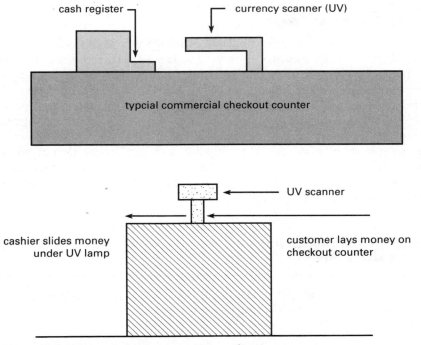

Figure 8-2. Discreet checkout counter application.

pany for handling the transaction, the losses sustained from denial of payment are large. Most businesses cannot withstand this additional drain on profits. Some retailers, especially supermarkets, go beyond the coupon value, offering "double face value" discounts. Although these types of offers, considered loss leaders to a store, create consumer advantages, the new crackdowns may end that benefit.

There are, of course, many unscrupulous businesses set up as scams to process coupons. In this chapter, I will focus on the legitimate retailer or wholesaler that falls victim to manufacturer's discount coupon rackets. Many coupon issuers have developed software and other techniques to identify color-copier coupons and other bogus reproductions. Since the reproduced coupons will have the same serial numbers and other encodings as the originals, they probably will be rejected, stamped void so they cannot be resubmitted, and returned to the store submitting them—*without* payment. Thus, if a retailer submits $5000 in coupons and half turn out to be bogus and returned without payment, that $2500 plus the processing fee will come out of the business's net profit.

Security Technique 94: Spot Bogus Discount Coupons

The ultraviolet system noted earlier can help you spot counterfeited manufacturer's discount coupons. First, you will need a genuine coupon to inspect visually and to study under a UV light. Next, pinpoint and list on a card elements of the coupon that stand out prominently. Under a UV lamp, the counterfeit, even if it was created on a copier from the original, will not appear the same as a genuine coupon. Giving every item scrutiny seems time-consuming. However, use of a scanning system, as described in Fig. 8-2, and implementation of other checks will go far in helping you detect bogus coupons. Contact the coupon clearinghouses and the manufacturers for additional tips on how to spot counterfeits.

Security Technique 95: Recognize Discount Coupon Scams

Organized coupon scams, at the local or national level, create significant losses for business. There are several types.

The Coupon Booklet Con

Consumers are often offered packets of local coupons, for hundreds of dollars worth of products and services, at a deep discount. Many of these promotions are legitimate. However, the key to recognizing illicit activities is to know you will not get something too cheaply. The scam artists play on greed. They offer you a deal that seems too good to be true—for example, $20 to $1000 worth of merchandise. The salesperson will show you examples of past successes that he or she had with the offer. However, these successes always took place in distant cities. The sample discount coupon booklet will also be phony. When approached, give the offer the benefit of a doubt and a bit of research.

If the salesperson shows you samples of other promotional efforts in other cities, ask to copy the information so you can talk to the supposed businesses in those cities. Ask the representative to come back in a day or two so you will have time to make the calls. If legitimate, he or she will return: otherwise, you will not see the scam artist again.

Make a couple of calls, one to the appropriate chamber of commerce and one to the better business bureau where the last successful effort allegedly took place. If the representative claims that the promotion took place in your town or city in past years, check to find out first if it was legitimate and then if it was successful. Chambers of commerce and better business bureaus keep records of consumer and business complaints. The company offering you the discount coupon deal might have legitimacy in a business sense, but its poor marketing record or failure to honor the booklets might create complaints.

Copy the name and address of the coupon company, then research it through the state agency that handles registry of corporations. Check with your own state to ensure that the corporation has been registered to do business there. Corporations, partnerships, and business entities other than sole proprietorships must register with the state. Many states have their corporate records within the Department of State and others within the Department of Commerce. Those offices can also tell you if the company has filed its annual reports; or you may find the state has dissolved the corporation for nonfiling. You can also discover the name of the person or persons who established the corporation as well as business and personal addresses and telephone numbers.

Copy the names of several businesses shown on the coupons, including those similar to yours. Call them and ask the owner or manager about the company making the sales pitch. If you established that the promotion coupons are legitimate, verify that they netted returns commensurate with the costs. If the deal is flawed or bogus, you'll find out quickly from the other businesses.

The Charity Club Con

Con artists also create consumer discount coupons for local businesses using the telephone book as their source. These people never contact the business directly. After creating thousands of booklets, the criminal enterprise goes to a variety of nonprofit organizations that regularly raise money for their own operations or charitable causes, and offers to sell them the booklets. The con artists may even pose as a nonprofit organization, using a "soundalike" name easily mistaken as legitimate. They offer to sell the booklets for a small percentage over cost, suggesting a deal that allows their victims to resell the booklets to local consumers and pocket the difference. For example, the make-believe charitable organization compiles the booklets for $3 each and sells them to churches and other nonprofit organizations for $10. They in turn sell them through their charity drives to locals for $20. The combined discount value of the book might amount to $1500 or more.

The problem begins when consumers try to redeem the discount coupons at local businesses, which have no knowledge of the coupons and refuse to honor them. The local charitable organizations get stuck with a bad name and huge losses. However, the local businesses also suffer immensely, because many purchasers believe that the businesses were somehow involved in the scam and made a killing from the sale of the discount coupon booklets. A business depending on local customers cannot afford to develop a bad name with the public. It can go bankrupt from lost business just as quickly as it can from direct crime.

The Chain Store Con

Similar scams target national chain operations such as fast-food or convenience stores which might be either company-operated or franchised. The coupons in the booklets appear legitimate, but often they are counterfeited from past legitimate promotions. When consumers who purchased the booklets present them at the store, usually at the end of the month, the store sends them to the company headquarters for credit. After a week or so, the company headquarters calls and tells the store that the coupons were unauthorized and counterfeit, and that no credit will be forthcoming. Sometimes the company and the store share the loss, but either way the business loses.

You can best protect your business by getting involved in the community through the local chamber of commerce, the better business bureau, the police, and a variety of organizations such as the local rotary club. Since most businesses can suffer losses of both money and credibility with coupon scams, the only way they can protect themselves is to spend a small amount of money collectively to publicize the scams. You can use a newsletter or even just a flyer and circulate it to all charitable and religious organizations, all businesses throughout the area, and all consumers. Local radio and television stations and newspapers will cooperate through free public service announcements warning people against discount coupon fraud. If you launch this type of campaign a couple of times a year, the con artists will look for communities that have not been alerted to their scams. In your own community, the campaign may help catch the defrauders in the act or enable the police to take action against the criminals.

Coupon Kings and Queens

A coupon king or queen is a householder who uses manufacturer's discount coupons and thrifty shopping habits to feed a family of five and keep everything running on $50 a week. Magazines, daily and Sunday newspapers, and direct-mail advertising inundate consumers with discount coupons issued by manufacturers. These coupons have become a second type of currency. However, they are often used in a variety of ways far different than originally intended.

Clubs and organizations have cropped up over the past 20 years that are involved in trading coupons. Businesses have been created to handle coupons. Newsletters and books tell consumers how to use coupons and spend less, and charity fund raisers sometimes base their activities on the collection and sale of coupons. Nearly any type of operation that involves genuine currency or checks will in some way include discount coupons. Although many businesses scrutinize checks, large bills, and credit cards, rarely will they check discount coupons. One key reason is that the coupon value usually ranges only from 10 cents to 75 cents. Discounts above these amounts are rare. Cashiers simply note the expiration date and honor the coupon.

One misdirected entrepreneur in Florida found the business of dealing in coupons lucrative, and grossed about $32 million over a 10-year period before being discovered and convicted. The "coupon king" stashed the money, served a few months in jail for fraud, and then disappeared to parts unknown. Some grocery store owners have discovered ways of creating enough income from dealing in real and illicit coupons that selling groceries has become secondary. The store serves as a front for their lucrative coupon operations.

Some householders have surpassed merely feeding the family and maintaining daily operations with discount coupons. One such householder became the "rebate queen." She created a sizable business in counterfeit proof-of-purchase bar codes using several names, and established over 50 post office boxes to collect the booty. She received a 2-month house arrest penalty for her enterprise.

Aside from the income that can be had from dealing in illicit discount coupons, there is little risk of being caught. It is even harder to prove the defrauder had fraudulent intent. Prosecutors usually are unwilling to take these cases to court because proving coupon fraud beyond a reasonable doubt requires a massive investigative effort, and jurors tend to be sympathetic to the defendant. It's hard for a jury to believe that anyone did any real harm with discount coupons. When convicted, coupon fraud offenders rarely receive significant punishment.

Security Technique 96: Watch for Cashiers Who "Moonlight"

If your business accepts manufacturer's discount coupons or if you issue your own coupons in local publications for promotional purposes, you may be tempting some employees to a secondary source of income. Most stores that sell Sunday papers clip only the header parts off the newspaper to obtain credit for returns to the company that places the papers in the store on consignment. The rest of the paper, including numerous coupon sheets, goes into the dumpster.

"Moonlighting" cashiers (or other employees) go through dumpsters on Sunday or Monday night to collect all the discarded manufacturer's discount coupons. Instead of using the coupons for the discounts offered, they use their position to convert the coupons into ready cash—your cash. Some of the coupons are obtained from within the store itself. Certain products have a peel-off discount at the time of purchase, and some brands of cigarettes periodically offer coupons worth $1 to $3 with larger purchases. ("Buy two and receive $2 off the regular price.")

The cashier "redeems" the coupons at the cash register. No shortage is noted because the coupon in itself represents money and the store gets an added percentage for handling and processing. There's no theft in a strict sense because you don't sustain a loss. However, this activity can create a cash flow problem, and if the cashier gets greedy and counterfeits coupons on a color copier, you might start finding clearinghouse rejections of the counterfeit coupons. Then you will sustain a loss.

A solution to this problem is to program your cash register to accept coupons only for products and merchandise already "rung-up" during a single transaction. The program should calculate a total for coupons entered into a specific register. The merchandise total can be compared against the coupons redeemed. When it is not possible to program the register, watch the flow of manufacturer's coupons from a specific cashier as compared with other cashiers. Show the coupons from the cashiers to the floor manager who supervises the restocking of shelves. The manager can usually tell you what's moving and what isn't. The manager should supply an accurate daily record of what goes from the stockroom to the shelves. This kind of double check is not only a good management tool but also protects you against losses from cashiers or others who attempt to convert discount coupons to cash.

Security Technique 97: Avoid the Coupon Fraud Trap

With increasing countermeasures taken by manufacturers and clearinghouses to detect coupon fraud, many offenders have begun looking for new approaches. You need to exercise caution to prevent falling into the coupon fraud trap.

Instead of opening store fronts to funnel coupons to the clearinghouses and manufacturers, criminals have started approaching legitimate businesses with seemingly lucrative deals. They might offer to sell coupons for 20 cents on a dollar, or by the pound. These operators will distance themselves from any hint of impropriety by claiming to represent some reputable organization. They research the targeted merchants carefully and alter their approach accordingly. For example, they tell the merchant that the proceeds above administrative costs will go to some charity or environmental cause. It sounds good, but even if the coupons are legitimate, it's illegal. The scam artists will use some genuine coupons but will counterfeit most of the ones they give or sell to a business.

One scheme in a western state involved a "retired professional fund raiser" who traveled the country contacting nonprofit, charitable, and environmental organizations that regularly needed funds for their activities. This "retired" fund raiser convinced one local charitable organization to participate in a "legal" way of raising large amounts of money through redeeming discount coupons. He told the organization that several manufacturers had authorized him to make the offer and that they would supply the coupons instead of cash. The money donated would come from advertising and public relations budgets.

The person's credentials appeared impressive. He claimed that after retiring, he had decided to do "volunteer work" across the country to keep active and apply his knowledge and experience to worthy causes. However, he said that the offer hinged on the merchants assisting in the fund-raising effort. The local charity, having a good reputation, agreed to aggressively recruit them. One more catch to the deal was that the participating merchants and the charity were to "contribute" a fraction of the coupon value to cover his "expenses" so he could go on to help another well-known charity organization raise money.

When the fund raiser had all the participants involved, he arranged for a truckload of coupons to arrive at the charity organization's headquarters. The coupons were neatly packaged and labeled, so it appeared that they had been sent by the respective manufacturers. The charity agreed to disperse them among the cooperative merchants. Within a few days, the fund raiser received their check for over $50,000 and left town. It was not until 2 months later, when a clearinghouse returned the majority of the coupons stamped "Void-Invalid," that the store owners realized they were counterfeit or expired. The police found the "retired" fund raiser to be a professional con artist whose true identity remained unknown. The participating merchants and the charitable organization were out the $50,000. In addition to the monetary loss and embarrassment they suffered, the retailers now had a questionable relationship with the discount coupon clearinghouse. With increased scrutiny of discount coupon transactions, you can avoid loss of money and reputation using this example as a "lesson learned."

Quick Reference Guide to Chapter 8: Detecting Counterfeit Currency

The best way to detect counterfeit currency by sight is to compare a genuine bill of the same denomination with the suspect bill. Look for the key elements described below. It is important to check systematically for differences rather than similarities.

- In genuine notes, the printed area of the back is always smaller than the printed area of the face. Hold a bill up to the light to see the difference.

- Look for differences in color or shade and variance in lines of design and quality of paper.

- Look for the red and blue fibers that are engrained in the paper of authentic currency.

- Check the portrait and oval background. A genuine portrait will appear lifelike and stand out distinctly from the fine, screenlike background. The photomechanical processes used by counterfeiters usually create a dullness in the portrait and an extra dark or light shading in the background.

- The check letter and faceplate number appear on the lower right corner of the bill, just above the denomination numeral. On authentic currency, both will appear clean, and the check letter will match the other check letter in the upper left corner of the bill.

- The Treasury seal and serial numbers will be clear and distinct on authentic currency. Most counterfeit bills have irregularities, and the seal's points will be missing or appear sawlike, uneven, or blunt.

- The Latin or English inscription inside a genuine seal is legible; on many counterfeit bills, it is not. (The English inscription was substituted for the Latin one beginning with the 1963B series of $1 bills. It was phased in to all bills afterward, completely replacing the Latin inscription after 1969.)

- The difficulty of reproducing the minute lines of the background of a genuine bill is so great that in counterfeit bills, these lines usually appear to be almost solid black or have numerous white spaces showing.

- Photoengraving is incapable of producing the same perfect line quality as the geometric lathe. The lines of the lacy design of a counterfeit bill, even under fairly low magnification, will appear spotty and look like a series of dots.

- Photographic reproduction is hampered by the several different kinds of letterings: some are solid black, others are partly black, and partly shaded, and still others, like the signature, are script facsimile.

- A counterfeit Treasury seal will lack the heavy outline of ink produced by the typographic method of printing that is used for the seal on genuine bills. Magnification will reveal the difference.

- The serial numbers on genuine bills are closely aligned, with the same spacing between prefix and suffix letters as between numbers. Counterfeits are not as evenly spaced and aligned. The letters and figures in the counterfeit may be of a different design, shape, and appearance from the original.

9

Ending the Shoplifting Problem

Shoplifting is one of the fastest-growing crimes in the country. Written accounts of shoplifting go back to the sixteenth century, when an Elizabethan chronicler described how professional troupes worked. Pilfering probably began when merchants first displayed their wares for sale. Specific methods of shoplifting appear to have changed little over the years. Techniques are influenced by the lifestyles of a particular society at any given time. Prevailing wardrobes, store architectures, displays of goods, and the nature of the shopping experience all affect how shoplifters operate.

A Time-Honored Profession

Shoplifters still employ time-honored methods, such as concealing items underneath their clothes and putting goods in false-bottomed cases or in well-worn shopping bags. Booster boxes—cartons and packages that appear sealed but have an opening for receiving "lifted" items—are still in use today.

Within the community of shoplifters, there has long been a minority of professionals, called *boosters* and *heels* in the trade, who make all or part of their living from this line of work. Exact losses from shoplifting nationwide are difficult to pinpoint. The mysterious disappearance of goods (inventory shrinkage) cannot always be traced to shoplifting. However, from known cases and a variety of well-researched and documented studies, shoplifting costs businesses several billion dollars a year. Many informed estimates top $40 billion. Items stolen by caught-in-the-act shoplifters are only a fraction of this total.

174

Some studies suggest that females and children shoplift more than males. However, I caution you against accepting this premise. It's based only on *apprehended* shoplifters and is therefore unreliable. For one thing, store owners and employees may be less likely to confront males. Females and children may present less of a physical threat to the apprehending owner or employee. The same applies to elderly shoplifters, who account for a significant number of offenders. Most store owners and employees would prefer not to accuse or apprehend elderly people.

One method of preventing the crime is to "catch" shoplifters in the act. In larger stores, this involves the use of house detectives. Unfortunately, sneaking around and apprehending a shoplifter who steals a few dollars' worth of goods has no effect on deterring the crime. While the store detective is busy pursuing a suspicious person, many others can steal at will and leave undetected. Professional groups of shoplifters will plant one or two "suspicious" types in a store to divert detectives while the others go about their illicit craft. Closed-circuit television and convex mirrors have equally little effect. Only an occasional amateur is caught, while others leave unnoticed.

Shoplifting may contribute heavily to your business problems or be your greatest loss. The good news is that you can control or eliminate shoplifting at your place of business. You can stop it by understanding shoplifters and their crime, by assessing your business's vulnerability on that basis, and finally by applying the preventive techniques suggested in this chapter.

Security Technique 98: Recognize the Common Types of Shoplifters

It's important to develop a clear understanding of shoplifting and of the types of people who threaten your business. Here are the common characteristics of shoplifters you will encounter at your place of business.

Amateur Adult Shoplifters

In most businesses, the amateur adult shoplifter is of primary concern. Typically, the amateur steals on a sudden temptation or impulse. Successes from initial thefts create stronger impulses and a greater psychological need to steal.

Occasionally, amateur shoplifters steal from a real or perceived need. The theft may not be for personal use. They may sell the items stolen to others for cash. Ironically, amateur shoplifters, if caught, will have enough money with them to pay for the items taken. They simply perceive the theft as an "economic" measure.

Most amateur shoplifters are self-conscious and will display symptoms of nervousness and uneasiness. Eventually, though, amateurs become as skilled as professionals, so do not rely on a preconceived notion that all amateurs are easy to spot.

Juvenile Shoplifters

Juvenile shoplifters can "nickel and dime" a business into ruin, especially in areas where young people flock into a business. However, knowing what they will take can help you develop countermeasures to eliminate opportunities.

Juvenile shoplifters follow certain patterns. They take small, perceived "luxury" items for their own use or to give to a friend. Remember that what you perceive as a "luxury" will not necessarily be the same as what juveniles consider desirable.

Often juveniles work in groups or gangs, distracting customers and store employees by noisy, disorderly behavior designed to cover their thefts. Many join the gang on a "dare" or "to belong," because they want acceptance from their peers.

Some juvenile shoplifters receive coaching by an adult, who often accompanies them into the store and pretends to be a customer. Occasionally, the juvenile and adult are related, or pretend to be related, to add credibility to their presence and avoid the suspicion of store employees.

Professional Shoplifters

Professional shoplifters make part or all of their living from stealing your business assets. They plan their work carefully. Much like burglars and armed robbery offenders, they "case" the targeted store as legitimate customers at least once before stealing. Many professionals make regular, legitimate trips into a store at certain times of the day to create a "familiarity" among store employees. They present themselves as polite, somewhat talkative types. All this throws off suspicion when they come back at another time to shoplift.

Professionals make a career commitment to shoplifting and often receive systematic training through an apprenticeship with seasoned veterans. They take pride in their skill and have neither a desire to reform nor a sense of remorse for their behavior. They often work in association with other professionals and may travel across the country, always maintaining an above-average lifestyle.

Professionals do not take many chances. They will desist if the time is not opportune and will "dump" merchandise, often openly, if they believe someone in the store has noticed their activities.

Professionals have a variety of techniques to lure a store detective into the open. Sometimes, a "respectable" member of the team will go to an employee and report having observed a "person" in aisle 7 pocketing several items. The suspicious person is the team's decoy, planted to distract store detectives so the rest of the team can "shop." Typically, the clerk calls for "number 4" on the store intercom to report to "aisle 7." Professional shoplifters are well aware of this security code and trigger it deliberately to expose house detectives. They then close in for the kill. The decoy takes nothing and moves quickly through the store with the detectives following. He or she watches for a signal from the team, then darts to the front of the store and goes outside. The decoy may then be stopped by a group of store detectives and detained for arrival of the police, who search the "suspect" and of course find nothing.

Another diversionary tactic involves having a team member faint while pushing a shopping cart with a few items in it.

Professional shoplifters, in business for the money, prefer valuable, salable merchandise. Many earn a net profit of several hundred to several thousand dollars a day. They may use specially crafted aids, such as booster boxes, coat-length pockets, and belt hooks. Most professionals steal in quantity and prefer a 3-day workweek to avoid overexposure.

Professionals steal for resale. They have fences or contacts, such as "bargain shops," that buy the items they steal. Often, professional shoplifters "steal to order." They may have a list of sizes and colors and may sell their stolen wares to "respectable" citizens at 50 percent off store cost. These citizens will deny any knowledge that the goods were stolen.

Senior Citizen Shoplifters

For professional shoplifters, the inevitability of growing old presents certain advantages. Senior citizen shoplifters succeed by virtue of their age. Most store employees picture them as their own grandparents or other elderly relatives rather than as professional thieves.

Many senior citizen shoplifters operate as couples who are or appear to be husband and wife. Some live and operate in large cities; many others operate out of a "Sunbelt" state where they maintain a retirement home. They travel the country in motor home caravans during tourist seasons or over the Christmas holidays. The motor vans offer them freedom of movement and a place to store their stolen wares, and also give them the appearance of being retired couples on vacation. In addition to shoplifting, these groups of senior citizens buy merchandise with stolen or counterfeit credit cards, forge or counterfeit traveler's checks, make counterfeit currency, and pass bogus personal checks.

Kleptomaniacs

Genuine cases of kleptomania are rare. Compulsive theft, true kleptomania, is comparable to being a "firebug." The kleptomaniac shoplifter is usually young, steals openly, and seems to "want someone to know he or she steals."

Kleptomaniac shoplifters may be male or female and come from every economic level. They take items without regard to value or use. They steal compulsively, often openly, and repeat the theft even after several apprehensions.

Drug Addict Shoplifters

Drug addicts turned shoplifters are the most dangerous type, because of their desperate need for money and their fear of imprisonment. They can become violent when confronted. Shoplifting and burglary are the most common means (studies suggest that females prefer shoplifting) of getting money for their "habit."

Drug addicts shoplift because of compelling physical need. They take long chances, often grabbing merchandise and making a quick getaway. When addicts steal they are usually at their lowest ebb, and their erratic behavior may enable you to identify them.

Vagrant and Alcoholic Shoplifters

Vagrant and alcoholic shoplifters also steal because of physical need, such as hunger or alcohol addiction.

Alcoholics are often under the influence when they shoplift and may use a "snatch and run" technique. Vagrants are more likely to be "just passing through" an area. They leave soon after stealing from several stores. Many have lengthy police records and may also be alcoholics.

You can spot vagrants and alcoholics easily by their appearance. Since they are not likely to have money, it is a fair guess that they are in your place of business to beg or steal.

Security Technique 99: Be
Aware of Shoplifting Methods

Except for those in desperate need of drugs or alcohol, shoplifters are usually outgoing people with a very presentable appearance. If you suspect them at all, it will be because of their emotions and actions. They modify and improve their methods with great ingenuity, often tuned to the level of alertness among employees. They are primarily concerned with obtaining and secreting items, avoiding detection, circumventing payment at registers or checkout areas, and departing from the premises.

There are two conditions that every shoplifter seeks:

- To find an area of the store that is unprotected. There the shoplifter can take an item and place it in a pocket or bag without notice.

- To determine when the store manager and employees become busy with other customers. Holiday shopping periods and rush hours are ideal times for the shoplifter to strike, especially in larger stores with more customers than clerks and fewer eyes on the merchandise.

Some of the more ingenious shoplifting techniques are detailed below.

Clothing Games

Loose-fitting clothing can conceal a large quantity of shoplifted items. Often hidden pockets or false linings are sewn into garments. Loose, tightly cuffed sleeves are especially good for concealment.

While trying something on "for effect," such as jewelry, a customer can unclasp the item and allow it to drop down the neckline. Patting or arranging the back of the hair can also "cover" dropping a small item down the neckline. A sweater or shirt tucked inside a suit jacket "to see how it looks" may be conveniently "forgotten" and worn out of the store. A tie or scarf can also be shoplifted in this manner. While stepping around the end of a counter to "see something," customers can steal expensive items from unlocked showcases.

The fitting room "shell game" is used to steal tight or closely fitting garments, which can be worn undetected under street clothes. Packages are rearranged to hide the addition of an article of clothing. Many shoplifters simply fail to return merchandise to the racks—for example, by getting more than the permitted number of try-on items without the knowledge of a clerk, and then returning only the permitted number while the clerk is busy with another customer.

Two customers in one dressing room, each with the allotted number of items, or one going back and forth to obtain garments for the other to "try on," can confuse salespeople about the number of garments taken and the number returned to the racks. The "difficult to please, hard to fit" customer who sends the salesperson back repeatedly for more garments also relies on this ruse.

Handling several articles as though for inspection, while reaching between customers, enables a shoplifter to ostentatiously return part of the merchandise and secrete the remainder. A variation is to get the clerk to display more stock than he or she can keep track of.

The Ticket Switch

A technique that is rising in popularity involves "switching tickets" on merchandise—that is, putting a ticket for a lesser amount on a higher-priced garment. A variation is to "mark down" the label in a forgery of the store's method of repricing items by hand. Ticket switching is difficult to prove. The shoplifter denies any responsibility for the wrong ticket being on the merchandise. This technique is popular in supermarkets that do not use bar code readers. Labels can be switched or pasted over genuine price labels.

Palming It Off

Palming is the simplest and most common method for removing small articles. It is difficult to detect, even when observed. The shoplifter places an open hand on a small article, squeezes the muscles of the hand to gain a grasp, and then lifts the object. However, the hand still appears empty. Palming is often aided by using a package, handkerchief, or gloves.

Child's Play

Children sometimes are trained for shoplifting. The accompanying adult discreetly points out the articles to be taken. If caught, the adult apologizes and "scolds" the child for theft. Also, adults may hide shoplifted items under clothing worn by children riding in shopping carts or within the toys that children are carrying. Baby carriages, and any personal items deposited in them, are convenient for hiding stolen articles.

Other methods of shoplifting include consumption of goods and collusion between customers and store employees. Remember that the confidence, dexterity, and ingenuity of accomplished shoplifters, if left unchecked, mean considerable net losses to your business.

Security Technique 100: Look for Shoplifting Accomplices

Earlier I mentioned that groups of people, amateur and professional, work together to create a "cover" for their shoplifting efforts. The following guidelines will help you detect a possible shoplifting scheme in progress. Remain alert to any disturbance and immediately give your attention to *all* customer activity in the store. Here are some methods used by shoplifters with accomplices:

- One person stands as a shield and acts as a lookout.

- One member of a couple engages the attention of a salesperson while the "husband" or "wife" strolls about the department, with a coat over his or her arm.

- High-value merchandise is moved to another rack remote from a salesperson's observation, where it is easier to pick up.

- One person transfers merchandise to a confederate. They may pass each other while pushing shopping carts, discreetly transfer items from one cart to another, and then move on in different directions.

- They create a disturbance or distraction: a family pushing a stroller with triplet babies, a rowdy group of teenagers, a person fainting, an elderly person falling. Disturbances serve not only to hide shoplifting but also to cloak thefts from cash registers as a cashier rushes to the aid of someone nearby.

- One confederate attracts suspicion and attention of store employees, requires much waiting on, or gets into an argument with the salesperson, while an associate gets away with valuable merchandise.

Security Technique 101: Assess Shoplifting Opportunities

My favorite advice on protecting your business assets applies to shoplifters as well: Eliminate opportunities and you will eliminate the crime. Several key elements create conditions conducive to shoplifting. After discovering these conditions, apply the countermeasures described below to eliminate opportunities. Most shoplifters, even the most amateurish, will recognize the futility of their efforts and leave you with your assets intact.

Look at your current floor plans and layout. Patron traffic flow, small rooms or partitioned areas, congested conditions, narrow aisles, cross aisles, and partly hidden or isolated displays create conditions conducive to shoplifting.

A lack of patron guidance and an absence of rules for shopping create needless confusion, congestion, and operational problems. Adequate warning signs or printed instructions help to control the admission of privately owned shopping carts, packages and bags, and baby strollers. Shopping carts must stay in the sales area. This eliminates the possibility that items on the lower shelf of the cart will go unrecorded at the checkout counter. If you permit patrons to wheel out shopping carts to a parking lot, caution cashiers to check the lower shelf of the carts. The best procedure at a checkout station is to place the bagged items into a separate cart, preventing patrons from going to the parking lot with the same cart used during shopping.

Shoplifting opportunities are greater when you have untrained, inexperienced, or indifferent employees. Through lack of training, interest in customers, or supervision, they do not observe the conduct of others. Insufficient or inefficient employees add to the problem. Shoplifting is the greatest threat when employees are absent and when coverage is low.

Go outside your place of business and reenter pretending that you are a would-be shoplifter. Look over the store as a possible target. How easy would it be to steal your business assets? You might be surprised at how easy it is. In your role as a prospective shoplifter, assess your business:

- What areas of the sales floor are most vulnerable to shoplifting?
- Does the arrangement of merchandise in vulnerable areas supply easy access?
- Are there areas that are not visible to employees?

Involve employees by discussing their past observations. Have they regularly found merchandise in parts of the store where it does not belong? Have they discovered empty cartons, small boxes, or other containers on shelves, in shopping carts, on floors, or in other locations? Employees can supply you with a wealth of information.

Evaluate your store's lighting, looking for shadowed or dark areas, or for lights that reduce your employees' ability to see throughout the store. Check the height of display cabinets to determine if they provide concealment for shoppers standing upright. Keep items of small size and high cost in protected cases and within easy view of an employee.

Ensure that you have the least number of operational doors consistent with safety or fire regulations. Your required emergency exit doors should have crash bars with alarms that sound when opened. Do you have the doors clearly marked?

Above all, your business policy should include these directives to employees:

- Wait on customers promptly.

- Avoid turning your back to a customer.

- Do not leave your section or department unattended, especially during busy times.

- Become especially attentive when groups of juveniles enter the store.

- Prevent children from loitering and handling merchandise.

- Observe people with baby carriages closely.

- Carefully observe people carrying large handbags, shopping bags, umbrellas, and folded newspapers.

- Stay alert to people who carry merchandise from one location to another.

- Do not allow merchandise to lie around a counter if it belongs elsewhere.

Security Technique 102: Use Layout and Atmosphere to Deter Shoplifters

After completing your vulnerability assessment, you can create effective countermeasures to end your shoplifting problems. Begin with the arrangement of your store. The layout should allow clear observation of customers by employees and yourself. Figure 9-1 shows a poor store layout encouraging shoplifting. Figure 9-2 shows a good layout that contributes to a sound shoplifting prevention program.

Note that in Fig. 9-1, cashiers have no view of the store, and employees working within the store cannot observe customers. Exit doors, including those to and within the stockroom, are not locked and are obscured from view, enabling shoplifters to leave the store easily. This layout has an abundance of areas that enable a shoplifter to hide, steal, and escape with little risk.

By contrast, the layout in Fig. 9-2 enables several points of observation by cashiers plus other employees working in the store. You can increase deterrents for the rear of the store with CCTV cameras and small screens at the checkout counters. This warns possible shoplifters that, besides employee observation, electronic detection creates a high risk for their activities.

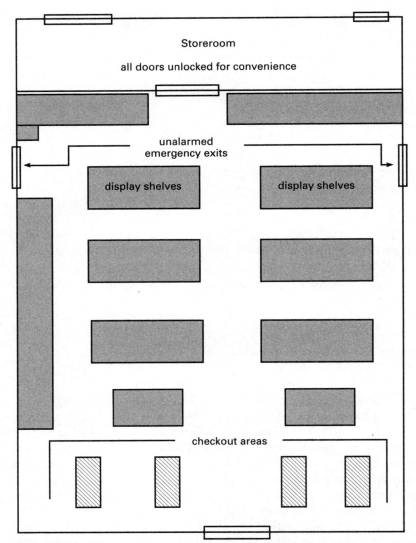

Figure 9-1. Example of a store layout that encourages shoplifting.

Keep the sales area neat and restock shelves regularly. Do not leave boxes sitting around on the sales floor, especially with merchandise in them. It is more difficult for shoplifters to steal from filled displays than from those that are half filled. Clear aisles and full shelves help employees observe customer activities. By walking along an aisle where a suspicious person has been observed, an employee can determine what is missing and then, under the guise of checking stock,

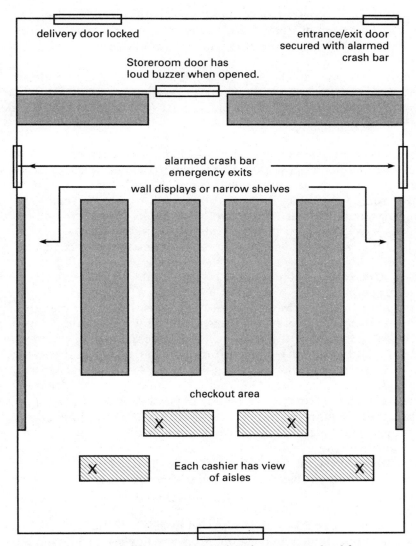

Figure 9-2. Example of a store layout that discourages shoplifting

observe what is in the patron's shopping cart or basket. The employee might carry a clipboard with invoices attached and hold a price marker. Shoplifters will know what is happening; honest customers will not even realize that the employee is watching them.

Ensure that every employee understands the importance of arranging display items so they can be easily observed and need minimal patron handling for adequate examination. Limit amounts of like items on display to permit easier observation and provide a continuous visual inventory of stock. Also, make an effort to remember faces. Shoplifters often return more frequently than other patrons. Relay a description of those suspected to other employees.

Honest patrons appreciate quick and efficient service. Shoplifters are wary of close attention and efficiency. They prefer employees who engage in needless conversation with each other or friends or who are "too busy" to help.

Closely inspect all items held by patrons while they show price markings and offer payment. Price tags may have been changed or switched, or container covers and caps may have been exchanged to show lower price markings. Whenever possible, cashiers should touch, lift, and examine merchandise rung up on cash registers. Weight and feeling may indicate that items have been added. Employees should not hesitate to examine suspicious items attentively.

Only authorized employees should handle sales, refunds, exchanges, and other transactions. Merchandise should be wrapped, tagged, or stapled properly, according to a policy that firmly establishes proof of payment.

Remember that an atmosphere of courteous and alert efficiency enhances your overall preventive effort. Employees should be well supervised and properly trained. They should be motivated to be neat, courteous, and efficient while performing their duties and serving patrons to the best of their ability. Such traits help not only to end shoplifting but to improve legitimate customer relations, thereby increasing the operational effectiveness of your business.

Often, wall or ceiling mirrors, closed-circuit television cameras (live and dummy), and alarm or warning devices on display cases make a wise investment and effectively deter shoplifters. Establish an effective daily inventory system, as noted in Chap. 6. This will also help you identify shoplifting problems and develop effective countermeasures.

Remember that the most effective and economic method of controlling patron shoplifting is through your employees. Effective human controls can never be adequately replaced by physical security devices or systems alone. Solicit suggestions from employees on methods to increase security. Evaluate their suggestions carefully. Effective ideas at the operational level often elude supervisors or managers, who are removed from the shoplifting problem.

The Return Merchandise Scam

If your business accepts returned merchandise, you need to create procedures that thwart crafty shoplifters. A method often used by professionals and aspiring professionals is to "shoplift" without actually removing items from your store.

After sustaining significant losses, a department store on the East Coast discovered this scam. The management of the store had decided that good security for screening returned items was to require customers to enter the store, pass by the checkout counters, and walk to a return counter located deeper in the store. According to the store's policy, customers could return items for cash or exchange only if the merchandise was in good condition and they had a valid receipt for it. However, to accommodate customers and ensure good relations, an unwritten policy made exceptions according to the prerogative of the cashier at the return counter. In some cases, a cash refund was allowed when the only verification was that the merchandise came from the store (i.e., the logo-embossed price tag was attached).

During a Christmas rush, the store was hit by a series of shoplifters who entered the store, selected merchandise that cost over $20, and then removed the merchandise from racks or displays. They walked to the return counter and asked for a refund, saying that they had lost their sales receipt. After verifying the price tags, the cashier paid the patron-shoplifters in cash. The year-end store inventory revealed that losses from "inventory shrinkage" totaled more than $500,000. Further examination discovered that a sizable amount of the loss involved refunds for merchandise that came directly from the store shelves to the refund counter.

Even if your policy is to demand a valid receipt, skilled shoplifters can get around this safeguard. Some will look for receipts outside the store, especially in and around trash cans where customers remove paid-for merchandise and discard the bag. Other shoplifters will buy several items in a store for personal use or for resale (a type of shoplifter's loss leader) and then use the receipt or a part of it to shoplift the same items. They take the items to the return counter, present the stolen receipt, and ask for a cash refund. The shoplifter will offer the excuse that the items do not fit, are not needed, or were bought by a friend who misunderstood his or her request.

Some stores tear off or remove price tags at the time of purchase, hoping to prevent a shoplifter from taking merchandise off the shelf and obtaining a refund. However, knowledgeable shoplifters will also remove the tags. They simply purchase items from the store to learn its method of tag removal and then duplicate the process on the items that they shoplift.

Yet another shoplifting technique involves returning items for *exchange*. The shoplifters find a receipt or make a purchase to get one. The scam artists later return to the store and shoplift the same items, placing them in a rumpled bag containing the valid receipt. They then ask to exchange the items for one or two other items of equal value. Later, they walk out with "laundered" stolen items—and the store's blessing.

Security Technique 103: Prevent Phony Refunds and Returns

To prevent refund and return merchandise scams, place your return section as near as possible to the entrance of the store and clearly post signs directing customers to it. Include warnings that no request for a refund or exchange will be honored if items are taken into the store without being cleared through a checkout counter. Potential shoplifters will then have to pass through a checkout counter for verification. If they bypass the refund area and carry refund items into the store, the risk of detection increases to unacceptable levels. When your store layout is effective and your employees are attentive, and when you have taken other security countermeasures, shoplifters who are planning to use the refund scam will have too great a fear that their activities are being observed.

Catching Shoplifters in the Act

Apprehending or accusing a person of shoplifting can become a legal nightmare for business owners. You need to develop a clear policy now. Do not wait until a shoplifter is caught. First, talk with the local police, the local prosecutor, and your business lawyer. If a suspected shoplifting case is handled incorrectly, the theft of an item worth $20 could cost you the entire business and more. To create a working draft of your policy, use the following guidelines. Remember to talk with the three agencies noted above *before finalizing any business policy for dealing with shoplifters.*

First, understand that neither you nor your employees have any authority other than that of a "citizen's arrest." The powers of a citizen's arrest are much weaker than those exercised by the police or other law enforcement agencies. A police officer has the authority to use force if needed to take a suspect into custody and also has the right to search the suspect for a dangerous weapon. Most police departments make searching an arrested person mandatory—to ensure the safety of the officer and others in the area, and of the suspect as well.

With few exceptions, state statutes prohibit store owners, managers, or employees who are making a citizen's arrest from using force or searching a suspect—even if they believe the person has a weapon. Nor can such people make a citizen's arrest using a gun, nightstick, or any other device to hold the suspect at the location until police arrive. The only exception is the right of all citizens to defend themselves when the situation will not permit withdrawing or safely escaping. Private citizens can protect themselves with force equal to the amount that is directed against them by another person.

Always ask yourself two questions before trying to apprehend a suspected shoplifter:

"Am I sure the suspect is shoplifting?" If you have any doubt, do not confront the person.

"Is there time to call the police for assistance?" Police officers are well equipped to handle the situation. With a citizen's arrest, difficulties and complications can arise.

Security Technique 104: Take Precautions in Apprehending Shoplifting Suspects

If you decide to apprehend a person for suspected shoplifting, keep three important factors in mind:

1. An employee must see the shoplifter remove merchandise from the shelf and hide the item on his or her person.
2. The same employee must have the shoplifter in sight continually while he or she is in the store.
3. The same employee must see the shoplifter leave the store without paying for the goods.

Some states have a "willful concealment" statute that enables the store employee to ensure the first two elements and then summon the police to make an arrest in the store. This regulation is separate from the shoplifting statute requiring that a suspect leave the premises before action can be taken. However, it has the same effect. Check with your local prosecutor and police department to determine your state's policy.

Always be wary when employees report that a person has shoplifted. A case of mistaken identity, a simple error in perception, or an employee overreaction can have devastating repercussions on the employee, you, and your business. If you do decide to apprehend a shoplifting suspect, be sure to have more than one employee present.

There are few successful prosecutions of shoplifters because of the difficulty of gathering evidence. Ensure that all your employees have a clear understanding of the rules:

Don'ts

1. Do not use force to detain a shoplifter.
2. Do not search a shoplifter for stolen merchandise.
3. Do not fight with a shoplifter.
4. Do not accuse the shoplifter of shoplifting.

Dos

1. Call the police immediately after discovering a shoplifter.
2. Detain the shoplifter through normal conversation until the police arrive.
3. Make any accusations of shoplifting after the police arrive.

Whatever policy you decide upon for dealing with shoplifters caught in the act, remember that with rare exceptions you can best eradicate shoplifting by eliminating the opportunities.

Security Technique 105: Use Countermeasures for Seemingly Honest Customers

One last, but important measure involves protecting your business not only against shoplifters but also against patrons who browse your store and do not steal outright, but instead try to underpay you for items. It's still theft if a customer can get through your checkout counter and leave the store with a receipt and stapled bag. This is a subtle form of stealing that merchants often overlook. These patrons alter the numbers on a price tag or replace it with a store tag from a less expensive item. Here are a few solutions for this type of theft.

Put two price tags on an item, one easily seen and one in an unobvious place where only cashiers and managers will look. Also, to help you guard this secret, have your cashiers handle items discreetly except when they discover that a customer has obviously changed a tag. Most people who find the second tag after they get home don't give it a second thought, concluding that the item was accidentally labeled twice by the store stocker.

If you staple your price tags to merchandise, use special staplers that produce a distinctive pattern. If the customer playing "switch the tags" tries to use a pocket stapler, the staples won't match. This will tip off the cashier to make a routine "price check."

If you use simple marking methods, don't write the prices in pencil. Customers can easily erase a price and change it. Use a rubber stamp or a pricing machine, or at least a marker pen.

Consider using tamper-proof gummed labels that rip apart when torn from the item. This type of price tag stops the switching game fast. Also, make certain the labels carry your store name and logo. Some ingenious underpaying customers buy their own gummed labels of the same generic type as yours, imprint them with a "more acceptable" price, and then attach the labels to merchandise while in your store. As they shop, they systematically remove your tags and replace them with their own.

Attach price tags to soft goods with a hard-to-break plastic filament. Go one step further and buy the filaments and tags with some type of marking that cannot be easily duplicated. As a final preventive measure, place some type of mark on the tags yourself when stocking.

Many department and clothing stores install electronic article surveillance (EAS) systems at their exits. If you're interested in such a system, make sure you choose one that not only sounds a loud alarm when unpaid items leave the store, but also verifies prices when a cashier misses the underpaying patron's scam.

Use care in accusing underpayers of theft. Although they may be guilty, you're on thin ice and in danger of a lawsuit, which the patron is likely to win. An underpayer may get cold feet or simply not have time to complete a scam. The person you catch might be an innocent customer who picked the item altered by the underpayer. The best approach is "detention." Have your cashier make the suspected underpayer wait for a price check while other customers stand around impatiently and stare. The discomfort alone should keep underpayers from coming around again. However, remain alert because the policy might also create a challenge for determined underpayers.

Quick Reference Guide to Chapter 9: The Shoplifter's Bag of Tricks

- Loose-fitting clothing can conceal a large number of items.
- Double-elastic waistbands create hidden pockets inside skirts or trousers.
- Hairdos can be arranged to hide small items.

- Rubber bands with a suction cup or hooks may be fastened inside a coat or jacket sleeve.

- Slits often create false pockets in jackets, skirts, or trousers.

- Wide skirts, baggy trousers, capes, and overcoats supply good hiding places for merchandise.

- A long belt with extra eyelets can be used to strap merchandise to the waist beneath outer garments.

- Knitting bags and briefcases can be used as pouches for small articles.

- A rubber band around bundles of ties, or stockings, or socks, can be used to hook items beneath outer garments.

- Clothespin snappers, wire hooks, or loops may be fastened under arms, or around a garter on the leg.

- Hats, gloves, sweaters, pocketbooks, and scarves may be worn out of your store but appear to be the customer's own property.

- Open briefcases and handbags, carried on the forearm or in shopping carts, are convenient places to drop pilfered items "accidentally."

- Shopping bags, bags or boxes from other stores, school books (popular for stealing CDs), and lunch boxes are perfect for concealing small items.

- Small items can easily be hidden in a soft drink or food container bought from a snack bar within the store.

- Umbrellas carried in loose or partly opened positions can hide small, lightweight items such as pens and jewelry. The shoplifter closes the umbrella when approaching checkout areas.

- Counterfeit bar codes, tailored to and placed on an item, can cause the store's bar code reader to reduce the item's price.

- Boots, raincoats, and other extra clothing worn during inclement weather offer additional places of concealment for the shoplifter.

- Boxes or paper sacks that contain merchandise paid for at other departments may be opened and then reclosed after pilfered items are added.

- Accumulated empty paper sacks can be filled with pilfered items and stapled with cash register tapes from prior purchases. These packages look like they are current, paid-for purchases and can be taken from the store without challenge.

- Folded magazines or newspapers, often hand-carried, may hide small, flexible items such as gloves and stockings, and other small articles in packages, bottles, or other containers.

- Coats or sweaters thrown down over merchandise can be picked up with the merchandise hidden inside.

10

Avoid Becoming a Victim of Business-to-Business Fraud

Those in business and the professions are targeted by practitioners of many deceptive and anticompetitive acts. Often, a company you have done business with for years suddenly begins victimizing your business through a variety of illicit schemes. Its profits and cash flow increase while the company continues to do business legitimately. Other allegedly legitimate businesses exist solely to victimize fellow businesses. This chapter examines several common practices employed by businesses that are out to victimize others. I discuss how to recognize the schemes, and provide solutions you can implement and actions you can take to prevent the huge business losses created by this type of fraud.

The Pervasive Gray Market

The gray market isn't product counterfeiting in a strict sense of the term, although it may include that crime. Nor is it black marketing (selling stolen goods), because the products are not stolen and they are displayed in legitimate retail or wholesale outlets. Often, the gray market doesn't involve statutory illegal activity. It does, however, create a subtle perversion of conveyance channels from a U.S. corporation to a

U.S. commercial operation or consumer. The *gray market,* in its most common sense, involves the merchandising or selling of goods domestically that are produced by a U.S. firm and exported to a foreign country. Unethical entrepreneurs have these goods moved back into the United States.

Security Technique 106: Understand the Mechanics of a Gray Market Scheme

Many U.S. corporations of all sizes produce a subgroup of their product lines for export purposes. Typical items include:

Cosmetics	Watches
Film	Cameras
Automobiles	Automobile parts
Electronic equipment	Soaps
Crystal	Consumables
Clothing	Appliances

Because foreign countries have less stringent marketing requirements than the United States, these exported products rarely meet U.S. standards of quality control, packaging, safety, purity, or other parameters that are set by regulatory and compliance agencies of the federal government. The products might also be obsolete, have no valid warranties, or be discontinued models. Because of relaxed foreign standards, a business in the United States can produce the products more cheaply and charge lower prices abroad than they can charge for comparable products (meeting U.S. quality standards) in the United States. The pricing often is so much lower that unscrupulous distributors, some of which may also do business with you legitimately, can create huge profits by purchasing the goods for export to a business that they own in a foreign country. Then they have their foreign business export the products back to them in the United States. Figure 10-1 shows you how the gray market operates through an otherwise legitimate distributor and how it may create major losses for your business.

Your business sustains losses from gray market operations in two ways: money and reputation. With some creative paperwork, the inferior products may never leave the United States at all. Instead, the paperwork makes the trip and it appears that the products were exported. The distributor then sells you these items as meeting U.S. standards. Remember, this type of distributor has legitimacy and per-

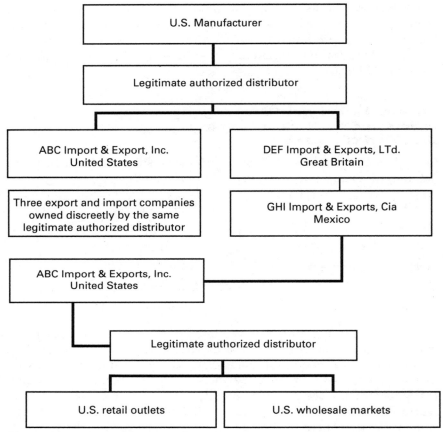

Figure 10-1. A typical gray market operation.

ceived integrity in the business community. Outside of its gray market schemes, it is honest in most of its dealings. The sale of a gray-marketed product, represented as a U.S. market product, can boomerang seriously on your business. When you complain to the distributor, you will probably receive only denials that the products were anything other than U.S. standard items. The major problem arises from increasingly adverse customer relations. The consumer buying a product from your store believes that he or she is getting a high-quality item, only to discover later that the particular product is clearly substandard.

Security Technique 107: Protect Your Business from the Gray Market

When you order products for your business, don't assume that they will be of good quality regardless of who supplies them, who manufactures them, or how long you have been doing business with the supplier. Even suppliers down the line from the manufacturer or distributor become victims of the gray market and sell to you without knowledge of it. The end buyer, usually a store, will be the first to unpack the product from sealed cartons. Before assuming the merchandise meets U.S. standards, examine it while it is in your stockroom, making sure that it does not carry telltale indicators of gray market goods.

Packaging.   Often the outside wrappings show damage or signs of tampering. Sometimes, the fine print on the package indicates that the product has been "imported" from a foreign country while also showing it as manufactured in the United States or in a foreign country for export to the United States. (The product manufactured in a foreign country for export to the United States must meet U.S. standards.)

Ingredients.   Gray market items may lack elements that enable a product to perform under U.S. regulatory standards. Because a manufacturer intended the product for a foreign market, it may contain prohibited ingredients or have substandard characteristics.

Labeling and Instructions.   Ingredients listed on a stick-on label (with the importer's name) placed over the manufacturer's printed information leave a question of authenticity and credibility. Also, look for a set of foreign language instructions or other information, including ingredients, without accompanying information in English.

Shipping and Warranties.   An "international warranty" does not apply in the domestic market. Your authorized distributor (who is technically responsible for the gray market) will not honor the warranty and will act innocent in this entire matter. To maintain a cloak of integrity, the distributor might even voice a complaint to the manufacturing company for allowing this type of gray marketeering.

Duplicate-Billing Schemes

The larger the business, the easier duplicate billing works. Established trade accounts are the biggest culprit. This "cash flow enhancement" often occurs in an otherwise legitimate business that needs money or in businesses headed by a management team that sees a lucrative way to increase profits.

In this scheme, a supplier sells widgets to your business, as it has for 12 years. When the bill arrives, you or your bookkeeper pays it without question because there is no reason to doubt this familiar business's legitimacy. The scheme often begins with a "test" double billing; however, because some businesses verify billings with invoices and receipt of shipment, the supplier may create a more complex system to confuse the issue. For example, the supplier might ship your business 1000 widgets, but send you a bill for 1500. The next shipment might have 900 widgets, with a bill for 1000. The rest of the shipments over a 3-month period, follow a similar pattern. However, the distributor always delivers a quantity that meets your sales or manufacturing requirements. After several shipments, an adjusted bill arrives showing the errors and adjusted amounts plus credits for payments. When the billing becomes complicated enough, you or your bookkeeper will pay the "adjusted" amount rather than spend long hours auditing this one account, taking inventories, and matching the figures to production or sales. These overbillings can amount to significant sums and in most cases are never detected. If a sharp bookkeeper does catch the overbilling, he or she will usually assume that an error was made instead of a fraud perpetrated, and the bookkeeper will send a letter or make a call. The supplier will normally send a letter of apology, but will state that its books are now being audited and that the company will get back to you when the audit is completed. If the supplier's auditors agree with your bookkeeper, a credit will be issued promptly. Meanwhile, the widgets shipments continue and the billings suddenly become accurate. After another couple of months, you send the supplier another letter about the earlier invoice differences and receive a reply that its auditors found a "slight error." This error is much less than you reported. The supplier agrees to continue to search for the problem. On the next bill, the supplier grants you a small, insignificant credit as a gesture of good will, hoping you will see that it has the right attitude and is trying to correct the problem. This scheme continues for months and usually involves periodic vague and confusing letters, even some that request detailed information. Because of your daily business pressures, you eventually allow the matter to fade. After the supplier has worked the same scheme among its lengthy list of customers, it will be your turn again a year or two later. This scheme might sound trivial and more trouble than it is worth to a supplier. However, it can produce several million dollars across the board when the supplier has a sizable number of accounts.

Security Technique 108: Take Action on Double or Triple Billing

All businesses, regardless of size, should have written policies. Even a sole proprietor needs policies on handling certain situations. However, as the size of your business increases, written policies become more important. One key policy you need to develop involves handling billings from your suppliers. Your policy should begin by calling for an exact count by those employees who receive goods. The invoice should be signed by the receiving employee and then given to you or the bookkeeper. Upon receipt of the supplier's invoice, pay only for what you receive. Do not rely on the supplier's billing or balancing the payment later (you make the adjustments, not the supplier). If the supplier tries to implement a double-billing scheme, the problem will soon stop when the supplier realizes that you pay only for what your business receives. Whenever you or our employees find discrepancies in a shipment, call the supplier immediately, and then follow up with a letter. Make sure you offer the supplier an opportunity to send out a representative to verify the shortage.

The Office Supply Scam

Office supply schemes often involve a sales representative calling on your business from a new "local" (legitimate) business. The sales representative informs you that the new company wants to supply your business with all its office and administrative supply needs at drastically reduced prices. He or she persuades you to give the company a chance, suggesting a small order at first and asking you to compare the quality and service with that of your current supplier. Because the business is new and local, you agree to a small "test" order. When the order arrives, you are impressed because the items are of the best quality and cost you considerably less than similar items from the supplier you've been using. There's usually a catch to the offer, however; because of the "drastically" reduced prices of high-quality supplies, you will have to pay on delivery.

The person delivering your order, accompanied by the same buoyant salesperson, insists that delivery services include helping you store your supplies. The company will not just drop off heavy cartons and leave the rest to you. This ploy allows the representative and delivery person to view your supply closet, get an idea of what you use, and determine your needs and the quality level of the items you buy. The salesperson will probably ask for another order, suggesting a few items that he or she viewed in your supply room. The representative tells you that the company can offer the lowest prices because the owners buy in vast quantities and will soon open stores across the country. If your business has branch offices or subsidiaries, you may soon begin to stock them with supplies from the new company, and certainly you will recommend the newcomer to other businesses.

After a few good orders, depending on the supplies you purchase, the scam begins. You continue receiving boxes that look like they contain high-quality items, but now the contents prove substandard.

Office supply scammers depend on volume and bait local businesses with genuine high quality at near wholesale prices. They do not make money for the first 3 to 6 months while they create confidence and build the size of the orders. They usually strike by announcing that the company has made a "tremendous deal" and can sell you everything you need for even greater savings. The suppliers rely now on getting huge orders, knowing the supplies will stay in wrappers and boxes for some time. After this big sales promotion, which includes shipping junk in high-quality wrappers and collecting on delivery, the company and its representatives vanish. Often, they start the same scheme under another name far away.

Another office supply scheme uses telemarketing methods. The office supply company calls your business and announces that it can supply your needs cheaper and faster because of low overhead—the company has no salesforce, stores, or employees. The caller asks specific, subtle questions to learn what kind of copier and other office equipment you have. He or she suggests a test order that you pay for on delivery and guarantees a full refund if you are not satisfied.

As in the example discussed earlier, the quality of the order is high and the prices are unmatched elsewhere. The telemarketing company gradually lures you into the big order and ships substandard items in high-quality containers and wrappings. When you finally discover the scam, the company telephones are no longer in service, and mail to the address listed on its advertising and invoices is returned as not deliverable.

Security Technique 109: Protect Your Business from Bait-and-Switch Supply Schemes

Whatever the nature of your business, don't accept any offers for office and administrative supplies that claim to give you prices far below those of legitimate suppliers. Every business needs to make a reasonable profit. If you shop around, you will find that prices do not vary significantly. When a deal seems too good to be true, it usually is.

Never buy office supplies in large quantities if you do not use them in large quantities. The "just in time" inventory concept can help you manage your business effectively and efficiently. Buy only what you need. If you pay for it at the time of purchase, go to a legitimate office supply store and buy it yourself. If you establish an account, do not agree to COD deliveries. Doing so removes your approval option and bargaining leverage when the items received do not meet your standards or are not what you ordered. (More on this bait-and-switch scam later in the chapter.)

If you delegate purchasing authority to an employee, make sure that you create a firm policy regarding purchases and be sure the employee follows it. In larger businesses, the slick salesperson will often try to persuade the purchaser through gifts (bribes) or kickbacks for big orders.

Unfair Methods of Competition

A variety of unfair methods of competition create massive losses for businesses by other businesses doing legitimate business. Court cases on this subject are too numerous and diverse to list in detail. The competitive practices most likely to injure your business (create losses) include:

- Using false or deceptive advertising

- Making unusually restrictive agreements, both horizontal and vertical

- Entering into conspiracies to fix prices

- Allocating markets, or preventing others from procuring goods

- Selling below cost to injure (create losses for) competitors

- Mislabeling products

- Engaging in gray and black market activities

Security Technique 110: Stop Losses from Deceptive Advertising

There's an ever-present question concerning the enforcement of laws regulating the advertising of products or services by businesses and advertising companies. Identification of clear deceptions is easy. However, subtle distortions are more difficult to pinpoint because even well-intentioned advertisers engage in puffery—exaggerating the advantages of a product or service. This practice is not illegal. Read any printed ad. Listen to any radio commercial. Watch any television commercial. You will invariably find examples of puffery. For example, a business might claim that its pens "write smoother." Smoother than what? Another example: "50 percent off while supplies last!" The question is, 50 percent off what? It is often very difficult to draw a line between what is meant only as puffery and what is deliberately deceptive.

In one area of advertising, however, the content is scrutinized closely. The Securities and Exchange Commission (SEC) has strict regulations against puffery in the advertising of securities. The SEC watches over the "tombstone ads" in the financial sections of newspapers to ensure that the advertisers confine themselves to basic statistical information about forthcoming stock issues. No discussion of a stock's merits is permitted.

The Federal Trade Commission (FTC) oversees advertising in other areas. Years ago, the FTC focused its efforts on fictitious prices and misuse of words like "guarantee." The FTC also attacked bait-and-switch advertising, in which a company advertises one product at a low price as bait to lure customers (similar to the office supply example above). The advertiser then announces that the product has sold out and tries to persuade the customer to purchase a more expensive product. Today, the FTC's objectives are much broader. Of particular interest is the "advertising substantiation" program, in which advertisers in widely varied fields are required to prove that the claims made in their ads are valid.

Although regulatory agencies serve as watchdogs, publicize their findings, and impose penalties that include corrective advertising, businesses can still be victimized by the deceptive advertising practices of others. An analogy to the courtroom is useful here. A shrewd lawyer often makes a clearly damaging statement that he or she knows will be objected to by an opponent, sustained by the judge, and stricken from the record. The problem is that the jurors *heard* the statement. The suggestion has been planted and stays in the backs of their minds. They will wonder about the truth of the lawyer's revealing statement, and may even perceive it as fact despite the court's instructions to ignore it.

The same damage happens to your business when a competitor practices deceptive advertising. Despite any excuses you might make or any government actions, the public and other businesses (and consumers) have *heard* the advertising and perceive it as true. For example, in September 1978 an FTC administrative law judge ordered American Home Products, then manufacturing Anacin,™ to correct its future advertising because earlier ads had claimed that taking Anacin relieved tension. Although the company had stopped making that claim by December 1973, evidence showed that customers continued to believe tension relief was an important attribute of Anacin. Holding that the image was likely to persist without a corrective message, the judge ordered that a $24 million ad campaign be launched to recant the original claim. The judge also barred the company from future advertising which stated that its Arthritis Pain Formula™ had "special" or "unusual" ingredients. Such ingredients are not unusual and are available in other products.

After years of litigation, an FTC judge took similar action against Listerine.™ He ordered the manufacturer, Warner-Lambert Corporation, to disclaim in its next $10 million worth of advertising all earlier assertions that the product would help prevent colds or sore throats, or lessen their severity.

Added legislation has enlarged the capacity and scope of the FTC since the passage of the Federal Trade Commission Act in 1914. The Wheeler-Lea Act of 1938 expressly bans "unfair or deceptive acts" in commerce. The Wool Products Labeling Act of 1939 protects manufacturers and consumers from the deliberate mislabeling of wool products. The Textile Fiber Products Identification Act of 1958 requires that clothing, rugs, and household textiles carry a generic or chemical description of fiber content. The Fair Packaging and Labeling Act of 1962 regulates packaging and labeling of food, drug, and cosmetic products.

Security Technique 111: Get the Attention of Government Agencies

If you are the victim of misleading advertisers, often the best you can do is put an end to the deceptive practice and try to demand "corrective" future advertising. You can also bring a damage suit, but false advertising claims are hard to prove. The most viable action you can take is to bring your claim to the attention of the FTC and other government regulatory agencies.

Food and Drug Administration

The FTC has a parallel agency, the Food and Drug Administration (FDA), which polices the labeling of foods, drugs, and cosmetics. The FTC is responsible for advertising in promotional media, while the FDA has power over claims that appear on the label or package. Added legislation in 1962 gave the FDA unlimited authority over prescription drugs. While the FDA's jurisdiction is confined to labels, its role is broader than that. If it finds that any claim in an advertisement is not supported by the information on the label, the FDA can proceed with a mislabeling action. The FDA has also cooperated with the food industry in developing standardized nutritional labeling systems for foods.

U.S. Postal Service

Closely allied to the FTC and FDA in its regulatory activity is the fraud staff of the U.S. Postal Service. Since 1872, U.S. postal fraud laws have supplied effective criminal and civil remedies against companies using the mail to deceive or defraud. Because fraud, from a legal standpoint, involves "intent to deceive," postal service cases ordinarily investigate misrepresentation that surpasses the "false advertising" handled by the FTC. Most problems have been in the food and drug fields. Postal inspectors refer the most serious cases of fraud to the Department of Justice; other cases are resolved through civil procedure hearings similar to those used by the FTC.

Federal Communications Commission

In broadcasting, the Federal Communications Commission (FCC) enforces rules regarding the types of products allowed in legal advertisements on broadcast media, the number and frequency of commercials allowed within a certain time, and what broadcast programs and commercials may or may not state or show. Under the Communications Act of 1934, the FCC received authority to regulate communications in "the public interest, convenience, and necessity." Through its licensing power, the FCC wields indirect control over broadcast advertising. Specific problem areas emphasized by the FCC are misleading demonstrations, physiological commercials considered in poor taste, and excessively long commercials.

Other Regulatory Agencies

Many other federal agencies have specialized roles related to advertising.

- The Department of Agriculture, under the Packers and Stockyards Act, plays a role similar to that of the FDA in regulating the labeling of meat products.

- The Environmental Protection Agency (EPA) regulates pesticide labels in the same way as the FDA regulates food labels.

- The Consumer Product Safety Agency asks advertisers to avoid any themes that may promote unsafe use of products.

- The Federal Deposit Insurance Corporation and the Federal Home Bank Board exercise close supervision over the advertising practices of banks and other financial institutions.

- The Truth-in-Lending Law and the Federal Reserve Board regulate installment credit advertising by banks.

- The Treasury Department's alcohol bureau enforces a rigid advertising code on alcoholic beverages that precludes many techniques used in other fields. These regulations specifically ban any attempt to impute therapeutic benefits to alcoholic beverages and any brand names which imply that a product originated in another country. For example, U.S. manufacturers may not name their product St. Petersburg Vodka or a similar label that implies Russian production. To avoid such problems, U.S. vodka producers use Russian-sounding names and adorn their labels with images that suggest the imperial crest of the czars.

- The Civil Aeronautics Board monitors the advertising of airlines and travel agents.

Violations of Antitrust Law

The antitrust movement was born in 1890, when Congress enacted the Sherman Antitrust Act (15 United States Code). *Trust* in this context means a combination of producers or sellers of a product that attempt to control prices and suppress competition. The intention of the Sherman Act is to preserve competition in the marketplace and to prevent further concentration of the vast wealth accumulated by a few corporations and persons.

The complexities and inner workings of antitrust law would take a book in themselves. Instead, this section will focus on the types of antitrust violations that are most likely to affect your business and suggest some remedies or solutions to keep your business from falling victim to unfair competition.

Antitrust law is both complex and controversial. It seeks to balance conflicting interests. On the one hand, it needs to protect the public from the power exercisable by all vast corporations. On the other hand, antitrust law does not seek to unduly hamper the capitalistic system of free enterprise. Antitrust legislation is intentionally broad and vague. It includes such phrases as "restraint of trade," "monopolization," "unfair methods of competition," and "conduct that tends substantially to lessen competition." This language has definite meaning only when courts interpret the statutes and apply them to the facts of particular cases.

Restraint of Trade

Section 1 of the Sherman Act declares illegal "every contract, combination in the form of trust or otherwise, or conspiracy, in restraint of trade or commerce among the several states, or with foreign nations." For Section 1 to be violated, there must be some combination or common action by two or more individuals or companies.

In interpreting the statute, the courts have equated "restraint of trade" with "restraint of competition." The interstate commerce requirement ("commerce among the several states") originally meant that the mere manufacture or production of goods was not interstate commerce even if the goods were destined for shipment from one state to another. Today, however, in order to satisfy the commerce requirement it is necessary to show clearly that the business or activity in question, even when purely intrastate, has a "substantial economic effect" on other states.

Rule of Reason and Per Se Doctrine

In a sense, every company or business contract restrains trade because it sets terms that stipulate that at least one commercial transaction will happen, and removes that transaction from competition. Obviously, this is not what Congress intended to prevent. Accordingly, in *Standard Oil Company* v. *United States* (1911),* the U.S. Supreme Court held that the Sherman Act's intention stays within the parameters by forbidding only "unreasonable" restraints of trade and that it does not limit the freedom of companies to enter into ordinary business contracts. However, according to the courts, certain types of business conduct do establish the condition of being unreasonable per se. This means that the conduct is itself unreasonable; proof that the conduct happened is enough to establish a violation of the Sherman Act. Examples of conduct that is *unreasonable per se* are price fixing and dividing market territories among competitors.

Price Fixing. Price fixing is any combination or agreement between or among competing businesses, formed for a specific purpose, that has

---

*221 US 1 55 L.Ed. 619, 31 S. Ct. 502.

the effect of raising, depressing, fixing, pegging, or stabilizing the price of goods in interstate commerce. For example, if several businesses form a trade association that sets the prices of their products and limits sales to a select list of jobbers, these businesses are engaged in illegal price fixing. There is no defense or recognized justification for price fixing—such as the need to end "ruinous competition" in the marketplace or the need to eliminate unstable prices that plague both producers and consumers.

The courts have been expansive in determining what constitutes price fixing, extending the concept to include agreements among competitors fixing minimum prices. In *Goldfarb* v. *Virginia State Bar* (1975),* the Supreme Court decided that a state bar association's establishment of minimum-fee schedules for attorney's services was illegal.

Division of Market Territories among Competitors. Any agreement among businesses performing similar services or dealing in similar products, so that the available market is divided and each business given a share, is illegal per se. As in the illustrations in the earlier price-fixing discussion, no justifications or defenses receive legal recognition by the courts. The rationale for this rule is that an agreement among competitors to divide the market for a particular product gives each an effective monopoly in its share of the market, because each firm has the power to fix prices on a particular product, even though competitive products are available.

Monopolies

Section 2 of the Sherman Act makes it a criminal offense to monopolize, attempt to monopolize, or combine or conspire to monopolize any part of interstate or foreign commerce. The courts have defined *monopolize* as acquiring or exercising the power to exclude competitors from a market or to control prices within relevant markets. By using this term instead of *monopoly*, the courts are defining an action (against which the state can act) rather than merely describing a status.

For example, Newspaper Company B distributes thousands of free copies in a geographic area in which Newspaper Company A operates. Company B makes this free distribution over a 4-month period, mostly on Wednesdays, the heavy grocery advertising day. Advertising creates 80 percent of Company A's revenue, and an important part of that revenue comes from grocery advertising. Company B has engaged in an illegal attempt to monopolize, because if Company B succeeds in driving Company A out of business, a strong and dangerous probability exists that Company B will achieve a monopolistic position in the

*421 US 773, 44 L.Ed. 2d 572, 95 S. Ct. 2004.

daily newspaper market and then be able to exploit that position to the disadvantage of the people and businesses in the area.

Relevant Market

At some inexact point, the subject of much antitrust litigation ("market power") becomes *monopoly power,* the power to control prices and to exclude competitors. To determine whether a business has monopoly power, it is necessary to define the "relevant market"—based on what creates (1) the relevant geographic market (for example, California, the West, or the nation) and (2) the relevant product markets (for example, the market for cellophane wrapping material or the inclusive market for flexible wrapping material). An antitrust defendant (a business charged with a violation of the antitrust laws) will argue that its business market is large, making its share of that market small. The business alleging the violation will argue that the defendant's business market is small, making its market share large.

A geographic market includes the area where the defendant and competing sellers market a product. If the product sells nationwide, the market is viewed as a national market. There may be regional submarkets where several local sellers compete with nationwide sellers; the market for beer is an example. In this case, the courts might view the submarket as the relevant market. Transportation cost is the principal factor limiting the size of a geographic market.

The determination of a product market often stems from consumer preferences and the extent that physically similar products can fulfill the same consumer need. In *United States* v. *E. I. Du Pont de Nemours & Company* (1956),* the so-called Cellophane Case, the U.S. Supreme Court held that goods found interchangeable by consumers for the same purposes create the same part of trade or commerce, monopolization of which may be illegal. The Court decided that Du Pont's relevant market was all flexible wrapping material, not merely cellophane wrapping material, stressing the functional interchangeability of these products. Under this rule, products like photocopiers would be perceived as interchangeable for antitrust purposes, even among brands using different mechanical processes.

Market Share

Courts often take the share of the market that an alleged monopolist has presently captured as the principal indicator of "monopoly

*351 US 377, 100 L.Ed. 1264, 76 S. Ct. 994.

power." Depending on the circumstances, as little as 75 percent of the market is sufficient to create monopoly power. A legal argument contends that market share creates an imperfect measure of market power, since it does not consider the availability of close substitutes for the product and reduces the ease with which new entrants, not presently selling in the market, can enter.

Tying Agreements

In a tying agreement, a person or corporation agrees to sell a product (the tying product) only if the buyer also purchases a different product (the tied product). For example, a cement company might agree to sell its premixed cement (the tying product) to a building contractor only if the contractor also buys cement mixers (the tied product) from the cement company.

The tying doctrine originated under Section 3 of the Clayton Act, which makes it illegal to sell or lease goods in interstate commerce if the purchaser or lessee does not use or deal in the goods of a competitor of the seller or lessor, if the effect is largely to lessen competition, or if it tends to create a monopoly in any line of commerce. There are three basic requirements for an illegal tie:

1. There must be separate tying and tied products.
2. The seller must have enough economic power to restrain competition appreciably in the tied amount of commerce.
3. The buyer must be obligated to purchase both products. If the two products are offered as a unit at a single price, there is no tie provided the buyer is free to take either product alone.

The tying product and the tied product can be related in a variety of ways. For example, they can be products used together in fixed proportions, such as nuts and bolts. The tied product may have a design requiring the use of the tying product, such as the data-processing programs used with a computer. Or tied and tying products, such as seed and fertilizer, may be used either together or separately.

Franchise Tying Agreements

The illegality of certain types of tying agreements has particular importance to the franchising industry. In *business format franchises*, the franchisee is granted the right to use, within a designated area, a franchisor's service mark, trade name, distinctive building structure, style of decor, and techniques of doing business. This type of franchising largely depends on linking the distinctive image to a standardized operation in the hope of maintaining high quality control. Because the commercial success of franchising depends on a nationwide uniform image, the courts tend to allow franchisors to require franchisees to use particular products to maintain consistent quality. However, if a franchisor insists that a franchisee buy *all* services, equipment, or supplies from the franchise, a potential tying problem exists.

Mergers and Acquisitions

A merger is one business's acquisition of the stock or assets of another in such a manner as to gain control of the business. The latter company is absorbed by the former and ceases to exist. Because mergers can lead to concentration of power in a few companies in an industry, they are closely regulated by antitrust laws. Section 7 of the Clayton Act, later amended, provides that a corporation cannot acquire the stock or assets of another corporation engaged in interstate commerce if, in any line of commerce in any section of the country, the effect of the acquisition may be to significantly lessen competition or tend to create a monopoly. The issues in merger cases usually relate to definitions of the product market (the "line of commerce") and the geographic market (the "section of the country"), and to the determination of the market shares of the corporations involved. Also considered are the status of the corporations as actual or potential competitors in the markets involved, and the probable effect of the merger.

A merger that violates Section 7 of the Clayton Act may also violate the Sherman Act. However, the showing of anticompetitive effect, essential to a successful prosecution, is much less rigorous under Clayton than under Sherman. Therefore, most merger cases are tried in court under Section 7 of the Clayton Act.

Interlocking Directorates

Section 8 of the Clayton Act prohibits a person from being on the board of directors of more than one corporation if these two conditions exist:

1. Any of the corporations has capital, surplus, and undivided profits of more than $1 million and is engaged in commerce. (The act excludes banks, banking associations, and common carriers subject to the Interstate Commerce Act.)

2. The corporations are or were competitors, so that the elimination of competition by agreement between them would create a violation of any provision of antitrust law.

Clayton Section 8 has been construed to forbid corporations from having the same director if an agreement between the corporations to fix prices or divide territories would violate Section 1 of the Sherman Act. By forbidding "interlocking directorates," Section 8 seeks to prevent violation of the antitrust laws by removing the opportunity or temptation to enter into illegal arrangements.

Price Discrimination

The Robinson-Patman Act makes it unlawful to discriminate, directly or indirectly, in price among different purchases of goods of like grade and quality under two conditions:

1. Any of the transactions involves commerce.
2. The effect of the discrimination tends largely to (a) lessen competition, (b) create a monopoly, or (c) injure, destroy, or prevent competition with any person who grants or knowingly receives the benefit of such a discrimination, or the customers of either.

The courts have decided that many types of conduct amounting to price discrimination are illegal under Robinson-Patman. For example, ABC Corporation is a large national company with a milk-processing plant in Louisville. ABC sells milk in surrounding areas at lower prices than in Louisville, where it has a large share of the market. Several local companies serve the surrounding areas and ABC Corporation is trying to get a share of that market. ABC Corporation's policy practices price discrimination, and violates the Robinson-Patman Act.

Both the seller who offers discriminatory prices and the preferred buyer who knowingly receives them are guilty of violating the act.

Like Grade or Quality.   Generally, physical differences in two products that affect their acceptability to buyers will prevent the products from being of "like grade or quality." However, differences in the selling brand name or label of the product alone do not create enough reason to justify price discrimination.

Defenses.   The Robinson-Patman Act contains nothing to prevent price differentials that make only "due allowance" for differences in "the cost of manufacture, sale, or delivery" resulting from the differing quantities, selling, or delivery methods. However, this defense will rarely have adequate legal weight for a defendant. For one thing, it is costly to compile the necessary proof. For another, according to the

Federal Trade Commission's interpretation of the defense, quantity discounts must be based on actual cost savings because of the quantity sold, not merely on a generalized policy that larger deliveries are automatically more economical.

A seller can also rebut a presumption of price discrimination by showing that its lower price originated, in good faith, to meet a competitor's equally low price; here, too, the defense is hard to prove.

Antitrust Remedies and Enforcement

The variety of lawsuits through which antitrust legislation is enforced is nearly as broad as the scope of the laws themselves, and each type of lawsuit has its own characteristics. The government can begin either a civil damages action or a criminal prosecution case against a suspected violator. Private persons can also seek damages through a lawsuit.

Civil Actions

The Antitrust Division of the U.S. Department of Justice has primary responsibility for enforcing the Sherman Act and the Clayton Act. The Antitrust Division investigates alleged antitrust violations from complaints received, often from businesses that believe themselves to be injured by a particular practice.

The Department of Justice has authority to begin a civil action to recover damages to the U.S. government resulting from violations of antitrust laws. One such violation is a price-fixing conspiracy that raises prices on goods sold to the government. In cases other than those seeking damages, the object of a civil antitrust proceeding by the Department of Justice is to secure a decree that will enforce the Sherman Act or Clayton Act by stopping or remedying violations. Antitrust decrees can enjoin whole categories of conduct and permit the Antitrust Division to investigate the subsequent business activities of the defendant.

Consent Decrees

Often, the government and an antitrust defendant agree to settle a case by consent decree, which stipulates that both parties suggest what remedy or relief the court should order, without the defendant acknowledging any guilt. However, the ultimate responsibility in fashioning an antitrust decree remains a judicial one. The court must determine what is or is not in the "public interest" and is not relieved of this responsibility simply because all parties to the proceeding (including the government) agree on a particular form of relief.

Private Antitrust Actions

Section 4 of the Clayton Act allows a private person who has sustained or will sustain business or property loss because of an antitrust violation, either of the Sherman Act or of the Clayton Act, to sue in a federal district court. A plaintiff who proves the violation of an antitrust law and resulting injury can recover treble damages (three times the proven damages) plus the costs of the action, including reasonable attorney's fees. A plaintiff who shows only threatened injury is entitled to injunctive relief, such as the issuance of an injunction prohibiting the illegal conduct.

In recent years, private antitrust lawsuits have increased. Most are brought as *class actions*—that is, all parties similarly injured by the antitrust defendant's conduct join as plaintiffs in the same action. Sometimes, actual individual damages may be very small (such as those stemming from a cheap consumer product). However, when action is filed by many plaintiffs in the class and the determination is multiplied by three under the treble-damages provision, the final award can become quite large.

Criminal Prosecutions

The Department of Justice can begin a criminal prosecution if (1) the case involves a "per se violation" of the Sherman Act and (2) there is proof of a knowing and willful violation of the law, such as a planned program of concealment. Most criminal antitrust prosecutions involve price fixing. As in other areas of criminal law, antitrust cases usually stem from a grand jury investigation. Subpoenas by the grand jury command the testimony of witnesses, the production of documents, or both.

Administrative Enforcement

The authority to enforce the Clayton Act on business firms subject to regulation by a particular federal regulatory agency is vested in that agency. For example, the Interstate Commerce Commission can enforce the Clayton Act on railroads and motor carriers, while the Civil Aeronautics Board can enforce the act on airlines.

Security Technique 112: Stay Attentive to Antitrust Developments

Our society continues to move toward more and more complex business relationships. Often, businesses participate in the growing complexities with less than honorable intentions. There's an erroneous view that the recognition of incipient crime is obvious. Some believe that the mark of a good businessperson is to "instinctively" recognize such problems. The result of such a position is to conclude that management needs no assistance in identifying incipient problems. Perhaps there are some legal problems that are obvious—those characterized in jurisprudence as *mala in se* (morally wrong in themselves). In large part, however, the legal structure within which a business operates and relies upon involves acts *mala prohibita* (made illegal by law)—a realm that is complex and difficult to understand.

Because of the expanding judicial construction of antitrust laws, an increasing number of antitrust cases have been filed by public and private litigants. The role of antitrust law has grown from that of maintaining commercial competition and a free market to that of monitoring business activities and producing social reform. It is essential that you develop an awareness of possible antitrust consequences of activities, especially when your business becomes a victim.

Quick Reference Guide to Chapter 10: Key Concepts in Business-to-Business Fraud

- The *gray market* is the merchandising of goods domestically that are produced by a U.S. firm for export to a foreign country. These goods do not meet the standards of their domestic counterparts and can cost your business not only profits but loss of reputation with customers.

- *Duplicate-billing schemes* are common among unscrupulous or hungry suppliers and can cost your business thousands of dollars annually. The remedy is to keep careful inventories and pay only for what you receive.

- *Bait-and-switch* supply schemes offer high-quality office products at exceptionally low prices. Once you agree to buy in quantity, perhaps even exceeding your inventory needs, you are shipped truckloads of junk in high-quality wrappers. Examine what you buy before payment, and buy only what you need.

- The *Sherman Antitrust Act* is the foundation of antitrust law. Section 1 prohibits contracts, combinations, or conspiracies in restraint of trade. Only "unreasonable" restraints of trade are prohibited, but certain practices, such as price fixing, are unreasonable per se. Section 2 makes illegal any act of monopolizing, attempt to monopolize, or combining or conspiring to monopolize any part of interstate commerce.

- Section 4 of the *Clayton Act* allows private individuals suffering business or property losses by any violation of the Sherman Act or Clayton Act to sue in federal court. A winning plaintiff can recover three times the damages sustained. Section 7 prohibits certain corporate mergers, Section 3 prohibits tying agreements, and Section 8 prohibits certain interlocking directorates of competing corporations.

- The *Robinson-Patman Act* prohibits illegal price discrimination if the effect may substantially lessen competition or if it tends to create a monopoly.

- Antitrust legislation enforcement stems from a wide variety of lawsuits, including civil or criminal proceedings by the Antitrust Division of the U.S. Department of Justice, private antitrust actions by individuals or classes of individuals, and suits by federal administrative agencies. Key terms in antitrust actions include:

Class action	Price fixing
Clayton Act	Relevant market
Consent decree	Restraint of trade
Interlocking directorates	Robinson-Patman Act
Market share	Sherman Antitrust Act
Monopoly	Treble damages
Per se doctrine	Trust
Price discrimination	Tying agreement

11
Safeguarding Your Business Information

According to a variety of studies, businesses lose between $2 billion and $4 billion a year to their competitors. The losses are not because a competitor has better management, better deals, or preeminent entrepreneurship. They are the result of a competitor's learning the business secrets of others: proprietary information. You might consider that you don't have any secrets worth stealing, but I assure you that your skills and business track record have significant value to comparable businesses.

Many business management advisers suggest that owners share their business knowledge with employees. In a perfect world, this advice is sound enough. But in the real world of business, not everyone has personal pride in his or her craft or integrity in personal dealings. The would-be experts contend that employees privy to inside information are motivated to work harder. If you have taken this advice, I can promise that your competitors will truly enjoy and benefit from your openness.

If I decided to start a new business to compete with yours, learning about the years of trial and error your business experienced would help me because I would not make the same errors and I would avoid the pitfalls. I would probably also cut deeply into your business volume and eventually might cause you to file for bankruptcy. My advice, as a business owner and a security professional, is to keep your business knowledge and sensitive information to yourself and let your employees have access to only what they need to know to perform their jobs.

This chapter examines how employees reveal information, how information brokers and competitors find out and use that data to their advantage, and how you can effectively protect your business. You worked hard and risked a great deal to get established. The details of how you managed to become successful are an important business asset and are just as valuable as the money and goods or services that sustain your daily business.

How Secure Is Your Business Information?

You don't need high-priced experts to analyze the security of your business information. Just take a look through your trash or randomly select a computer, tap a few keys, and you will see how easily anyone can access information from the business's databases. If your trash and the computer reveal more than they should, it's time to step up your efforts to protect the valuable inside information. If you think your business has no secrets to protect, you're not really competing with similar businesses.

Each business has its security nuances. However, I have developed reliable general techniques that will protect your information assets. You do not have to invest in costly high-tech devices. There are some devices that you might consider later, ones that have a cost-effective advantage in specific situations. For now, let's look at the important first steps to take.

Security Technique 113: Don't Give Away Business Assets by Telephone

Ironically, valuable and confidential business information may be innocently revealed to your competitors through telephone conversations. We have become addicted to telephones. They supply us with a way of conducting business quickly and with no restrictions on distance. Telephones help us sell products and services, make proposals and appointments, get information quickly, and close deals in distant places. However, all our telephone traffic has also made us willing to answer almost any question by phone. This is how you and especially your employees can easily give away valuable information to a competitor. Remember, your first problem with a caller is knowing his or her true identity. Unless you know a person well enough to recognize the voice, you depend on the caller's integrity—and that can be a big mistake.

One of my clients discovered that a competitor consistently had advantageous information about his company's operations and products. The client asked me to find out why. Company management thought that "industrial spies" were bugging the offices or operating inside the company, and despite the fact that the business spent vast amounts on security, data continued to be leaked. While reviewing the company's document security in its purchasing department, I overheard an employee talking to someone on the telephone, and discovered the problem.

The employee, one of the company's buyers, answered his phone and readily supplied confidential information. After he finished with the call, I asked him what the call was about. He said that Sam Smith in the company's research and development section wanted to know where the company had bought one of the components for the new widget machine it planned to manufacture and market later in the year. The buyer gave the caller the information without hesitation.

When I asked the buyer if he knew "Sam Smith" personally, the buyer shrugged and admitted he did not, but quickly added that the company had several hundred employees and that he knew only a few personally. He confidently took out a company roster from a desk drawer and ran his finger down to the research and development section. A surprised look came over his face. The name "Sam Smith" did not appear on the roster of employees, but the buyer said he could be new, since the list was a month old. I decided to check and was not surprised to learn from the personnel department that no "Sam Smith" had ever worked for the company. Since we know that this caller had no legitimate reason for requesting the information supplied, we safely assumed that the call came from a competitor or someone hired by a competing company.

With this discovery in mind, I began talking with a variety of company employees and found that they readily gave out information to callers who identified themselves as employees of the company. Although the company had a strict policy about revealing confidential information over the telephone, no one thought to remind employees that people anywhere could call and identify themselves as company employees.

Management moved quickly to develop a new telephone policy for this company. The last I heard, competitors were no longer preempting its marketing efforts.

No matter what size your business is, someone will always have an interest in finding out how you enjoy success and will benefit from your time, money, and hard work. The best advice I can give you is not to supply confidential information over the telephone, regardless of whom the caller claims to represent.

Security Technique 114: Limit Access to Your Business Information

It is important to limit access to your information as much as possible. Employees cannot tell what they do not know. Some employees will need access to information even when it has high value to a competitor. Assuming you use discretion in hiring and placing these employees, you still need to keep in mind that they work for a salary and have no personal stake in your business.

The crime of embezzling and selling proprietary business information or using it to begin a competing business has gained in popularity among employees in recent years. Although we would like to think that everyone works hard for his or her money and has undivided loyalty, we also know from studies that about 90 percent of employees work only for their salary and do only what they need to do. Within this high percentage, an equally high number will steal your assets without remorse if you supply them with the opportunity to do so. Your information has value, perhaps more than any other business asset, depending on your operations.

You can avoid losses from information assets (primarily from lost business) by an attentive system of limiting access. The U.S. government found that one of the best ways of restricting information and controlling information leaks is to enforce a rule of "need to know." If employees "need to know or have access," you must allow them to do their jobs. If they do not have a need to know, restrict their access to information assets.

Security Technique 115: Classify Your Proprietary Information

You cannot expect your employees to know what is important and what is not, unless you tell them. Developing an information classification system will help you control access to it. For example, you can assign a "classification level" to each employee that corresponds with levels of confidentiality. You might consider the following guidelines in your business.

Nonconfidential Proprietary Information

First, determine what information would come under the classification of nonconfidential. What would your competitors learn if they come into possession of certain data? If your answer is "nothing" or "not much," then you can safely classify it as nonconfidential (N). You have no need to restrict access to this type of information. On your employee roster, if your business has many employees, you can place an (N) as a suffix alongside each name.

Confidential Proprietary Information

The classification of confidential information begins with the access control of "need to know." Disclosure of this information will have important effects on your business, its customers, clients, or employees. Confidential information (C) might include such data as proposals, rosters, estimates, customer lists, purchasing data, and personnel records. You should create a controlling register for this classification. For example, when the document or information comes into existence, make sure that anyone taking it from your files has a clear need to see the information and has been authorized to do so. Be sure that employees sign for sensitive information documents and files on a register that includes the time and date that they receive the information, and when they return it. To best control this information, appoint a records custodian who is responsible for these controls and for employee adherence to them. In smaller businesses, you or your manager might personally control the information made available to employees.

"Close Hold" Proprietary Information

The exact title you place on the most sensitive information depends on your preference. I personally prefer the label "close hold" to "secret," because it sounds less like the language of government or a spy movie, and seems more appropriate for the business world. You should ensure that this business information asset (CH) has strict limits. Only those employees and others who have a clear "need to know" should be allowed access. You should create a controlling register for "close hold" information, as noted above.

The Perils of Public Places

Discussing your business in any public place pleases information brokers, who carefully research where company employees and executives or owners go for lunch, have afterwork cocktails, and meet for other social gatherings. The reason they go to such lengths to find out this information is that they want to be there to listen and record the conversations. Businesspeople "talk shop" with one another because that is what they have most in common. As a World War II slogan suggested, "Loose lips sink ships." I would advise you to consider that your own conversation might "sink" your business.

Controlling this type of information leak calls for employee awareness. Employees must be alert any time they gather in a public place, especially those who have access to confidential information, that someone might be listening to or recording their conversations. We tend to be creatures of habit. For example, if you go to a restaurant or local pub regularly, you will probably go to the same table, stool, or booth whenever possible. The late J. Edgar Hoover, director of the Federal Bureau of Investigation, went to the same restaurant each night of the working week with Clyde Tolson, his deputy director. They sat at the same table for nearly 40 years. How easy it would have been to install bugs to listen in on some interesting conversations that took place around that table. Inform and train your employees so that they will not talk about confidential business in public places.

Security Technique 116: Keep Your Business Travel and Meeting Plans Private

If you or your employees travel and conduct meetings as part of your business operations, ensure that they stay within the "confidential" or "close hold" classification. Your competitors or would-be competitors can develop valuable insights from this information, including your marketing opportunities, strategic partners, new suppliers, new clients, and major business deals in progress. They might use this information to preempt your operations. For example, if a competing business learns you have prepared a proposal for a big business deal with another company or client, the competition might call and undercut your offer, thereby getting the contract.

Often, travel and meetings seem trivial but can reveal valuable information to competitors. They may keep tabs on you or a key employee simply by calling a receptionist and asking for you by name. If the receptionist says that you are out of town, the competitor may have someone else call who poses as an important client, insisting that he or she will deal only with you and must know how to reach you. Chances are, the receptionist or secretary will be afraid not to tell the caller your location and telephone number. You need strict procedures for this type of situation. No information should be divulged. The receptionist should inform the caller that you will be contacted and will return the call. If the person is legitimate, he or she will accept the arrangement. If not, the caller will hang up.

Security Technique 117: Do Not File Too Much Information Outside the Business

Most business owners think little of telling government agencies, other business-reporting organizations, or even banks about their business operations. Although you need to remain honest with government agencies, check with an attorney who is familiar with business filings and give only the least amount of information necessary under the law.

Some organizations collect business information but do not always compile a business profile from it. Most banks, for example, can legally divulge "business" accounts and transactions and often sell that information to a variety of reporting organizations. That information might include your average balances, deposit records, and outstanding loans. A competitor could find this data useful in a general assessment of your business operations. Dun & Bradstreet also compiles reports for its own purposes.

You can protect your financial information best by having "asset management" accounts with a variety of investment-oriented companies that will not sell or report your information and will maintain minimal banking activity. You might also consider banking alternatives, such as money market and other bank-managed investment accounts, that will help you prevent competitors from learning your true financial prowess. Ambitious competitors will strengthen their own positions by acquiring enough capital to come on strong in marketing or to undercut your prices significantly. These measures could cause a devastating drop in your sales income, especially if your capital position is weak and you cannot respond.

Security Technique 118: Safeguard Information at Training Locations

Many businesses have locations scattered about geographically, each a small business within itself. Collectively, they form a large company.

Chain operations are an example, as are services and other types of businesses. Training and business meetings for these companies usually occur at or near the smaller business site, and often take place in easily accessible motel or hotel meeting rooms. These can become a treasure trove of information for competitors. They are easy to find through inside information or local announcements. Adopt the same controls and classifications at these locations as you do at your central business.

Internal Security Measures

Internal security begins by educating employees. Spend some time with your employees discussing the importance of safeguarding your business information. Start with how it affects profits and losses and their jobs, salaries, and benefits. Review the following ways a competitor might draw out information.

- *Flattery.*   An information seeker might play to an employee's ego, lauding his or her experience, expertise, or some other vulnerable point of vanity.

- *Downplaying the importance of information.*   The information seeker might downplay the importance of the information by telling your employee that it has no value beyond personal curiosity.

- *Bluffing.*   Information seekers might convince an employee that certain information is routine and there is nothing wrong with giving it out. Or they might claim to already know the information and then trick the employee by allowing him or her to "correct" them as they describe the information.

- *Misdirection.*   The information seeker may start a casual conversation with your employee about insensitive areas of the business and discreetly guide the employee into proprietary areas.

Once you have educated your employees carefully, take a good, hard look at your business premises and internal affairs.

Security Technique 119:
*Lock It Up!*_____

Lock up documents and sensitive information, and make sure that the keys to your filing cabinets do not open *all* filing cabinets. Most office filing cabinet locks will open with the same key and therefore supply minimal security. Anyone can go to an office supply store and obtain a file cabinet key that provides access to sensitive information that you believe is secure.

To safeguard your files without changing locks, consider installing some hasps and locking bars to back the filing cabinet lock with a good combination or key padlock. A professional will still be able to open it. In most businesses, however, it is nonprofessionals—employees, cleaning crews, and burglars—who pose the greatest threat of information theft.

Figure 11-1 shows ways of hardening your business filing cabinets, including adding locks. These are especially important safeguards for ensuring that information does not fall into the hands of competitors.

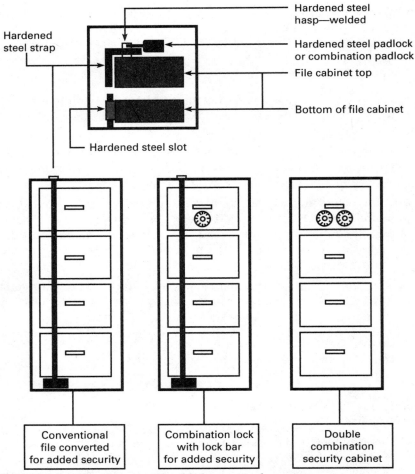

Figure 11-1. File cabinets with added security elements

Security Technique 120:
Safeguard Your Computer
System, Software, and Files

Few businesses today operate without computers of some type. The information you put into your computer will probably have significant value to a competitor. If you understand the fallacy of computer systems, no doubt you back up everything by copying the file and its updates onto floppy disks. An employee or anyone who might have access to your computer can steal the disks or easily copy the information onto floppy disks. Some businesses have lost their entire computer system and disk library to burglars, who then sell the computer and information together or separately to an information broker who has no scruples. (More on computer crime later in the chapter.)

Most computer sales centers have a variety of measures for securing your computer against theft, such as providing ways to prevent starting it up without a special key or other device. You can also install special software to prevent unauthorized access to information on a hard drive, and you should always lock up your floppy disk backups and program software in a safe or other secure container that will prevent access or theft.

Newer computer systems offer removable hard drives that you can lock in a safe when the computer is not in use. This can be a wise investment, especially when you have your business records in the computer database.

Security Technique 121:
Shred Your Business
Information

Inexpensive, portable shredders are a valuable security addition. Consider a model that fits on a standard plastic wastebasket, as found in most offices. Put one on each wastebasket in your business; you cannot go wrong. Instead of crumpling, tearing, or just tossing whole documents into the trash, employees can feed them through the wastebasket shredder in about the same amount of time. When you use this technique, you gain the advantage of knowing nothing important will go out in the trash and fall into the hands of competitors. Some information brokers make a sizable amount of income from rummaging through business dumpsters each night. They bag all the wastepaper and sort through it later for documents of interest to their clients, your competitors. Often, cleaning crews bag wastepaper and make the job even easier for the information scavenger.

Security Technique 122: Have Employees Sign Nondisclosure Agreements

You do not always need a nondisclosure agreement to bring a lawsuit against an employee for divulging your business secrets and other information. However, requiring such an agreement sends a strong signal to employees about how seriously you value business information. Ask your attorney to create a document that spells out to employees the consequences of their theft of business information. Be sure that the agreement stays in effect for a reasonable time *after* an employee quits, retires, or is fired. Otherwise, a person could simply join the business to collect information and then sell it after leaving without fear of prosecution.

Security Technique 123: Watch Sales Pitches and Advertising Brochures

Salespeople can become so enthusiastic about a product or service that they inadvertently give away valuable information during their sales pitches. Also, the people they are selling to may have a connection to your competitor. Another sensitive information leak stems from advertising brochures or letters about your business. Your marketing personnel or a contracted advertising agency may innocently print information you would prefer remained within the business. Here, too, employee education techniques are important.

Security Technique 124: Beware of the Dangers of Fax Machines

Show me a business today and I will show you at least one fax machine. These wonderful aids to transmitting information fast have created a whole new way of doing business. They have also provided a way for information thieves to work quickly and directly, stealing business information almost without a trace, often under your watchful eye. For example, an employee can take a "close hold" document and fax it directly to a contact at your competitor's place of business or to a broker who pays for the tidbit. Many businesses have developed elaborate security for their sensitive information, including rigid rules and security checks at the exits, to ensure that no employee walks out with original or copied confidential information. Unfortunately, these businesses forgot the dangers of fax machines. When using fax machines, the employees may look "squeaky clean honest" but all the while may be sending your classified information to a competitor. Also, most businesses leave their fax machines operational 24 hours a day to receive incoming fax messages. They forget that the janitor, for example, could find information and fax it out with less chance of discovery than a regular employee has during business hours. Employees who have access to confidential files can also come into the office during the evening hours or on weekends (under the pretext of catching up on work) and fax the information with no fear of detection.

The Vulnerability of Mobile Offices

A growing trend among businesspeople is to use sophisticated high-technology equipment, formerly restricted to a place of business, in their cars or home. Now, we increasingly see cellular telephones, portable computers and printers, modems for transmitting information from the portable computer through the cellular phone, and portable fax machines—all in people's cars. A businessperson can keep in constant touch, compiling, sending, receiving, even conferencing while traveling by car along a turnpike. Much of this technology is now in use on aircraft and trains, and in motel rooms converted into offices for one night.

All this is good news for your business operations. However, it is even better news for your competitors, especially when they know where you (or your valued employees) are at any given point. For example, with a few pieces of easily operated equipment, the information collector can simply follow you around, at a respectable distance, and find out the transmission frequencies of your cellular telephone from which the computer and fax sends and receives. The collector can then easily receive all the information going and coming from your car, some of which may be "close hold" or "confidential." Although the process will probably become illegal, it will still go on—much like exceeding the speed limit, rolling through a stop sign when there's not a car in sight, or throwing a scrap of paper out of a car window.

The dangerous part of "mobile" business information collection is that you do not know it's going on and there's no way you can find out. Unlike the conventional telephone bug or tap, which experts can detect and which installed high-tech equipment can counteract, the cellular car phone works off radio or microwave transmissions to or from a nearby receiver that relays through conventional systems. Once your transmission frequency is known, usually through an inexpensive scanner available at any electronics store, the information spy simply tunes the cellular phone, fax, and computer modem to yours. When you send information, the spy too receives, and when you receive information, so does the spy. All this occurs without interfering with your system.

Another danger with mobile business operations is that they penetrate deeply into your information systems. The information collector can not only obtain the information that you transmit or receive, or record both sides of your cellular telephone conversations, but also determine the entry code into your other computers, such as the mainframe that you contact through the mobile modem. Once your entry code and password are known, the spy can tap into that source and download or copy a massive amount of business information.

Finally, mobile equipment, programs, and information pose a security threat. Even if your car has an excellent alarm system that alerts you and others to a break-in, most professionals will risk triggering your alarm and will escape with your equipment and information before anyone has a chance to react. They may also just drive up quietly with a tow truck, remove the car to a remote location, and then steal your business information and equipment. Professional spies can glean data and then sell it on the black market to brokers who are not particular about where it came from. If you remove your equipment and put it in your motel room for the night, make certain you do not leave it there unguarded: Motel rooms are often easier to break into than cars.

Security Technique 125: Confound the Eavesdroppers

You need to become aware of the vulnerability of your mobile communications and impose safeguards when transmitting and receiving proprietary business information. Communications experts and developers of mobile equipment continue to bring out new attachments that scramble and encrypt information in order to confound eavesdroppers. However, be forewarned: The equipment will probably cost you more than you want to spend. If you need to send and receive a large amount of sensitive information, it might be cost-effective in terms of protecting business profits that could be lost to your competitors. If you want to save money, consider keeping your cellular telephone conversations coded and restricted to nonclassified business. Rely on overnight mail for hard-copy transmissions. Whatever you decide to use in your place of business, or in your mobile office, do not become a victim of the theft of your hard-won ideas, plans, strategies, and work.

Information Embezzlement and Computer Crime

When I mention the word "embezzlement," most people don't think of business or proprietary information, but instead think of money. The threat of embezzlement does involve money, but you also need to look upon your business information as money and assets, because it can be equally valuable. Embezzlement of any business asset, be it money, commodities, or information, usually involves five factors:

- Embezzlers of business assets plan their activities carefully and have access to what they seek.
- They conduct their embezzlement scheme systematically.
- They conceal their crime through whatever means necessary to avoid discovery.
- They create a situation that is hard to prove or prosecute if they are discovered.
- They participate only in schemes that supply significant financial gain.

The embezzler of business proprietary information has a variety of financially based motives, although it is not always readily apparent how they will reap any benefit. In the past, embezzlers used methods that today we might consider "primitive." Most contemporary embezzlement occurs through computers.

The following examples of computerized embezzlement show the scope, sophistication, and seriousness of this crime.

Check-Writing Fraud

A small check-writing service used computers to write and stamp-sign the payroll checks of large corporations. The corporate clients had thousands upon thousands of employees scattered about the country. The service company owners had no idea how computers worked but they knew the results the computer achieved, so they hired programmers to operate them. One programmer instructed the computer to find the average amount of the checks issued and then to write checks for that average amount, addressed to false names and locations. Later, the programmer cashed these checks. He also programmed the computer so that whenever it printed out a list of recipients' names and corresponding check amounts, it would automatically delete the false names for each corporation. The firm sent that list to the corporate client.

Since there is normally a regular turnover in a large corporation, especially among clerical employees, the balance on the list rarely corresponded exactly to the amount on the corporate payroll. However, because both balances were within 10 percent of the actual payroll, as the programmer assured everyone it would be, the corporate client never raised any question with the check-writing firm.

Even after the programmer decided to change jobs, he continued to receive payroll checks each week from different companies. Because the instructions to falsify the lists were so deeply embedded in the computer, it took another skilled programmer weeks to uncover the scheme. However, because of the technical problems, a delay in the indictment, and a tip from a friend who heard about the investigation, the original programmer vanished with millions of dollars. He is probably living under a false name in some country that has no extradition agreement.

Fake Insurance Policies

A notorious case of computer crime involved the Equity Funding Corporation. The case demonstrates that embezzlers and other corporate criminals can use computer printouts to support almost any claim they want.

Equity Funding was essentially an insurance company that, according to its records, was growing rapidly. The financial assets of the company nearly doubled in a short time and showed no sign of slowing down. Investors on Wall Street noticed this growth and quickly bought available corporate stock, causing the price to soar.

In reality, the firm was growing by fabricating insurance policies (on computer) for nonexistent people. The task was easy for the programmers and operators. A supervisor in the computer service department developed a special tape containing a complete list of nonexistent policyholders.

Whenever anyone asked for an inventory of insurance policies, the computer would list the real policyholders along with those that had been faked. There was no way for someone not privileged to the scheme to identify the fake policyholders. Only three or four top executives of the firm had knowledge of the plan. Whenever the company wanted to show growth, all the computer supervisor did was to add a few more nonexistent policyholders to the tape. The computer automatically credited the firm's financial records with the new policies. Whenever an outside firm, such as a bank, wanted to examine the financial condition of the company, it was presented with a computer printout that reflected the fake policies.

As the law required, outside auditors spot-checked policies. An auditor would randomly choose policies from the computer and ask the company to supply the documentation to support it. Sometimes, the auditor would send a letter to various policyholders asking for written verification that a policy did exist and was in force.

Equity Funding's plan worked until one of the computer supervisors inadvertently forgot to remove the tape containing the fake policies from the computer system. When the auditors started to select fake policies from the tape, executives who also were part of the scam began to write policies in nonexistent names to support the computer's claims. However, the executives could not write the policies as fast as the auditors were requesting the information. Soon, the auditors caught onto the scheme and called in the federal government to investigate.

Bogus Deposit Slips

A small bank in a suburban New Jersey community, just outside New York City, became the victim of a computer-literate thief from outside the company. A college student had access to a computer coding machine used to print such things as account numbers on checking-account deposit slips. The student opened an account at the bank, took a stack of deposit slips from the bank's courtesy desk, and coded them with his account number. He then returned the slips to the bottom of the deposit slip pile at the bank. Whenever a customer used one of those deposit slips, the teller would deposit the amount into the account listed on the slip. Instead of the money going into the customer's account, the money went to the student's account.

Before long, the bank received complaints from customers whose checking and savings accounts were not credited with the proper funds. The bank agreed that a mistake had happened but could not find the problem within its system. Only after a bank officer deposited money in a personal account using a blank deposit slip with the student's coded account number did the bank discover the scheme. An audit of the student's account confirmed that a "misposting" had taken place.

The bank and law enforcement officials unsuccessfully tried to prosecute the student for illegal manipulation of bank funds. The bank had the blank coded deposit slips, but there was nothing to link the student to them. The only proof the bank could offer was the improper transfer of funds. However, such an error could have happened without the student's knowledge. The bank decided not to pursue the case if the student allowed the "misposted" funds to be removed from his account. Claiming he knew nothing about these deposits, the student agreed.

Security Technique 126:
Understand How Information
Embezzlers Use Computers

In the past, only major corporations used computerized systems as part of their asset management programs. Today nearly every business relies on computers to compile and maintain business records and information. As this trend continues, new thieves or embezzlers of business assets also emerge. They can be any employee or any contractor, because most businesspeople today are computer literate. The special and highly accurate nature of computer accounting challenges the computer criminal's ability to outsmart the machine and your system.

For most business owners and managers, a technical gap exists between supplying the computer with raw data and receiving the material organized by the computer in return. Most businesspeople cannot instruct the computer on how to organize the material that comes from programs, which are usually prepackaged and available through computer supply centers or designed by an expert computer programmer to meet the specific system's need. If you choose the latter route, or hire an outside accountant to keep books on your computer, you are essentially supplying a second party with control over the information fed into the computer, over how the computer organizes that information, and ultimately over the form in which the computer returns this information to you.

Ideally, of course, a business does its own computerized accounting in house. However, if yours is like most businesses today, you have to entrust at least part of this important responsibility to others. If you cannot trust the people you allow to control your finances, commodities, and information on computer, you will have two major problems at the outset. First, the dollars and cents of your business operation might be handed to an information embezzler or other type of criminal. Second, you and most police agencies are neither equipped nor trained to uncover illegal manipulations created by the dishonest employee. Briefly, you may know the person is stealing from your business, but you do not have the technical skill to show the courts how it is being done.

Security Technique 127:
Determine Your
Vulnerability to
Computerized Crime

Computer crime can happen in many ways. The person controlling your business assets with a computer can create fake records, creatively code computer cards that give the computer misleading information, and instruct the computer to "steal" funds. Or the computer thief may attack the records of a business. Many businesses, small and large alike, have their payroll, inventory, sales, and other important marketing information in a computer file. Some businesses can support a complete in-house computer system; others use the services of an outside or time-sharing company. On the surface, detailed information about a business may seem worthless to outsiders. Remember, however, that all businesses exist within a competitive climate. Whether yours is a local liquor store or a company servicing corporate clients, you will have competitors trying to acquire your customers.

Many facets affect a business, from cash flow to the number of items sold in a given category. For example, a business owner or manager who knows that the competition cannot drop prices below a certain level without losing money can build up enough of a reserve fund to launch a 1- or 2-month campaign offering lower prices. If over a long period the other business cannot meet the low price, it will usually lose out to the competitor. The important consideration for the aggressive competitor is knowing how low a price to charge and how long to maintain it. This information will be available in the computer records of other businesses.

Smaller business operations are vulnerable to information embezzlement not only from competitors but also from suppliers. The smaller business usually cannot afford an in-house computer and will turn to an outside service. The outside firm will give each client a code to "open" its file in the computer. This arrangement could tempt one of your employees to sell the code to a competitor, giving it complete access to your business information. Your competitor can then determine the financial position and strength of your business.

Your suppliers also have an interest in detailed information about your business. Some suppliers will obtain inside information about you with a computer. This information will advise your supplier of the quantity of merchandise you can safely buy and will reveal your required profit margin. The supplier can then estimate the price at which you will buy the merchandise—for a marginal but acceptable profit for your business. You must give your computer files, whether in house or farmed out to others, the same care and accountability that you give to your money.

Security Technique 128: Know the Characteristics of a Computer Embezzler

What circumstances might turn a seemingly reliable employee into an embezzler? Dozens of psychological studies profiling the typical criminal have generally excluded those involved in purely economic crimes, such as computer swindle.

First, embezzlement means larceny (theft). Normally, an embezzler works alone and violates the trust of an employer or client. Often, management complacency supplies the opportunity to commit the crime, which may begin as a small experiment. When the experiment proves successful, the embezzler continues the scheme on a larger scale, digging deep into your assets. If the embezzler does not become too greedy, he or she may never be discovered. Even if the embezzler is found, most business owners will not pursue the matter, especially if the embezzler resigns and makes satisfactory restitution. A business does not want competitors, suppliers, bankers, and customers to know of its internal problems, especially problems involving dishonest employees. Some have even given favorable references to dismissed embezzlers to avoid revealing the true reason for their leaving the firm.

In some types of businesses, successful embezzlement schemes require collusion among employees. Usually, the collusion is confined to one area of the business and involves a small number of participants. For example, the use of remote terminal stations in distributed data-processing systems can contribute to the collusion among employees who handle different functions. A warehouse manager who enters shipping information directly into an integrated database may work in collusion with a bookkeeper in accounts receivable. The bookkeeper arranges false shipments without posting the corresponding invoice information into the financial files. The materials so shipped can become diverted for private sale. Elaborate schemes of this type are commonplace in the annals of computer crime.

The primary motivation for the computer criminal is financial gain. Aspiring to *la dolce vita* and seeking solvency after personal loss are two common reasons for computer theft. The violator tries through some form of embezzlement to solve his or her financial problems. Occasionally, an employee will embezzle because of perceived mistreatment by an employer. Then the scheme becomes a way of "getting even."

Embezzlers always rationalize their actions to appear devoid of criminal intent. Some employees may try to penetrate a computer security system just because it exists. If caught, they will contend that the system challenged their technical ingenuity and they needed to demonstrate their prowess. However, such employees are as much of a threat as the financial embezzler, because they will try to penetrate the most sensitive, best-guarded computer files—and that can cause disclosure of your confidential information. They may even commit theft if their interest shifts from playing a game to obtaining tangible results for their efforts.

Security Technique 129: Solve Computerized Embezzlement Problems

Unless you run a huge corporation that uses mainframe computer systems requiring a staff of experts, you *can* operate your own in-house systems. That's the first step to gaining control of your business information, commodities, and money. However, use caution in operating this system and in selecting the people who have access to it.

Whether you use an in-house or outside computer system, be sure to install a program that records who accessed the system and what information they requested, including the time and date. If you are using an outside firm, this log will deter someone from selling an entry code. In house, it prevents or discloses any unauthorized person who is trying to learn the code and enter it surreptitiously. A record of the date and time makes a difference as well, because an unauthorized person will have access to the computer during nonbusiness hours. Of course, an authorized person could access the information routinely during business hours, thereby avoiding suspicion. However, assuming that the computer programmer or other authorized party is innocent, the time and date will narrow the list of suspects to a janitor, an employee, or even a burglar. It is an excellent investigative tool and has the added benefit of revealing the information that was requested and received. That information may even help you identify the competitor who is receiving the data and may lead you to an employee who is on your competitor's payroll.

An embezzler stealing your business's money is easier to trace than one who is selling information. However, do not make any assumptions. As in the example of the check-writing firm, had the programmer's company practiced the right auditing procedures, the person could not have perpetrated that scheme in so many companies for so long. Whether you operate a huge corporation or a smaller one, establish every possible safeguard and routinely double-check everything. Whenever possible, do it yourself or at least learn enough yourself so that you can validate what others do for you.

Quick Reference Guide to Chapter 11: Safeguarding Computerized Business Information

Safeguard computers that control your business assets by using these "right" techniques:

- Operate an in-house system if possible.

- Lock up everything.

- Control access to your system.

- Know who accesses the system and when and what they receive.

- If you must hire outsiders to computerize your information and other assets, divide the work among several people so at least you will cut your losses.

- If you become a victim of a computer criminal, do not rely on the police, courts, or juries to help you. They do not have the computer expertise, the time, or the inclination to help a business that is ripped off. Juries rarely understand the technical side of the problem and often side with the defendant against a business that is prosecuting.

- Trust employees and others when they appear honest, but always verify their activities.

12
Preventing the Diversion of Business Cargo

All businesses sustain losses from cargo diversion, although you might not recognize it in your own operation. Even if you don't ship commodities from your place of business, you do receive them. When an incoming shipment of merchandise or products (or raw materials for manufacturing) has shortages, chances are it sustained some type of cargo pilferage along the way. Certainly, you don't have to pay for items you don't receive. However, not receiving all that you ordered may create a larger loss to your business because of lost sales or the need to slow down production.

Another business loss you might sustain from cargo diversion (theft of entire shipments) or pilferage (partial theft of shipments) relates to your suppliers and carriers. When they sustain direct cargo losses, someone must pay for their write-offs and increased insurance rates. Their prices for merchandise, products, or raw materials will inevitably increase. You want to make a reasonable profit, but you also have to compete with other businesses. So you cannot always pass along supplier or transportation price increases directly to your customers. Most often, you will have to settle for increases in the prices of products or merchandise that split the difference between yourself and the customer.

Finally, if you ship products, merchandise, or raw materials *from* your place of business, or feature a customer delivery service that uses owned or leased trucks, you might sustain direct losses through cargo diversion or pilferage. According to the U.S. Department of Transportation, business losses across the country from various categories of cargo theft run into several billion dollars a year.

Protective Management of Intransit Cargo

In this chapter, I address the common problems of shipping and receiving cargo, explain how and why pilferage and diversion happen, and discuss how you can better protect your business cargo. I also examine how you can help prevent losses created from receiving less than you order.

Much of your success in protective management of intransit cargo relies on selecting the mode of transportation that provides efficient delivery while also offering minimal opportunities for theft. In a key section of this chapter, I summarize the common modes of transportation, including their advantages and disadvantages. This concise security assessment will help you determine the best methods of shipping or handling commodities, both nationally and internationally.

Finally, if your business involves operating a carrier, terminal, warehouse, depot, or other type of cargo-handling facility, you will find helpful techniques to control cargo diversion and pilferage. Once again: Eliminate the opportunity and you will eliminate the crime!

Security Technique 130: Put an End to Shipping and Receiving Theft

Opportunities for stealing cargo abound in most shipping and receiving operations, regardless of the size or nature of the business. This type of business theft, like all others, calls for protective management controls that can be modified and expanded to meet the threats effectively. Here are the crucial "starting points" for your business.

Whenever possible, limit the hours of receiving shipments to certain times and post a schedule on your receiving dock. This technique not only has a labor cost-saving benefit but also enables you to establish better control over possible cargo theft.

Do not allow one employee to handle the three functions of receiving, stock management, and delivery. While this consolidation of responsibilities might save money, it also creates opportunities for an employee to commit cargo theft. Cargo-handling employees may have an opportunity to take part in a collusive pilferage or diversion plan with truckers and delivery people.

Whenever practical, separate the physical premises of each specific operation, such as receiving, inventory, and delivery. To avoid collusion, place the employees managing these receiving and shipping functions under separate department or management supervisors. Keep receiving and shipping doors closed and secured when they are not in direct use. Never allow your cargo to remain unattended on a shipping or receiving dock.

Ensure that your vendors or carriers know who in your business has authorization to receive or ship cargo. Never allow your receiving staff to verify an incoming shipment solely by the packing slip, carrier manifest, or invoice. There may be significant discrepancies between these documents and the actual shipment. Make sure that no incoming shipment of goods is commingled with existing or verified stock.

Ensure that your receiving employee makes an attentive physical examination and counts the merchandise before signing for the shipment. When you receive cargo that's verifiable by weight, check the weight of the total shipment or applicable components before accepting the order.

One of my institutional clients regularly ordered "large" eggs for delivery three times a week. Although the shipping containers (and invoices) claimed to hold "large" eggs, they actually held "medium" eggs. The institution receiving clerks now weigh the egg cartons and compare them with the known weight standard for large eggs.

Have your receiving staff immediately report damaged or opened goods to the delivering carrier. The report should be made during delivery and inspection, and in the presence of the delivering truck driver or other transportation representative. When an employee finds a defective container in an incoming shipment, he or she should clearly mark the container as "damaged," "opened," or "not accepted." The carrier representative should sign the invoice or other receiving document, attesting to the validity of the claim. Your receiving clerk should also immediately send copies of the invoice (with the deficiencies noted) to your purchasing and accounts payable departments.

Railroad boxcar or freight company truck and trailer shipments generally have serial-numbered metal seals attached to the doors. The number of the affixed seal is written on the bill of lading. Thieves often pilfer cargo from trains and trucks and replace the seal with one having a different serial number. If you receive cargo from these sources, have your employees verify that the seals are intact at the time of delivery and that the serial numbers on the intact seals match those on the bills of lading. When the numbers do not agree (although it might be an administrative error), the shipment needs special attention. It is likely that some theft of cargo occurred while the merchandise was in transit.

Security Technique 131: Create Warehousing, Inventory, and Stockroom Controls

Like the vault that holds a bank's cash reserve, the warehouse or stockroom of your business supplies a reservoir of wealth. Stock and warehousing operations have a track record of considerable cargo theft problems. Here are some of the most common problems, along with ways to end them.

Establish a perpetual inventory system. Almost all methods of eliminating or controlling commodity shrinkage in storage begin with daily or almost daily inventories of in-stock items. I am sure you are shuddering at the mention of daily inventories and can tell me dozens of reasons why doing so is impractical or impossible. However, I have helped establish inventory systems in hundreds of business operations, including warehouses and terminals, and I can attest to the fact that regular inventories are invariably beneficial. Whenever a business facility establishes and routinely uses a perpetual inventory system, it experiences *a dramatic reduction or elimination of theft.*

A simple and often overlooked technique for controlling inventory involves making a "total" count only once. Each later count establishes only "what is not there," instead of what is (another total count). For example, if you have a warehouse, stockroom, or other storage facility, segment the cargo or merchandise storage space into area A, area B, and so on. When you place boxes, containers, pallets, or items in area A, conduct a "total" inventory count and arrange them in a stack of, say, 50 boxes. When you conduct your inventory the next day, knowing the arrangement contained 50 boxes yesterday, you count the spaces left by the removed merchandise, instead of recounting all the boxes.

Let's assume someone removed 5 boxes from area A to include them in an outgoing shipment. With this system of inventory, you can quickly detect 5 missing boxes from the arrangement and note on your inventory form 45 boxes now present in area A. After you have taken your inventory, it is an easy task to account for the shortages each day with a check against outgoing documentation. Most businesses now use computer systems to speed the verification of incoming and outgoing cargo.

Figure 12-1 shows the arrangement of cargo in a storage or marshaling facility (such as a freight terminal). Figure 12-2 shows an easy way to create an inventory count sheet.

You need to follow up on your perpetual inventory system with periodic selective counts of certain items in stock. You should use an executive, manager, or senior supervisor for this process. To supply effective verification *and* a deterrent effect, make sure that the periodic verification count is unannounced. The selective count (of a certain

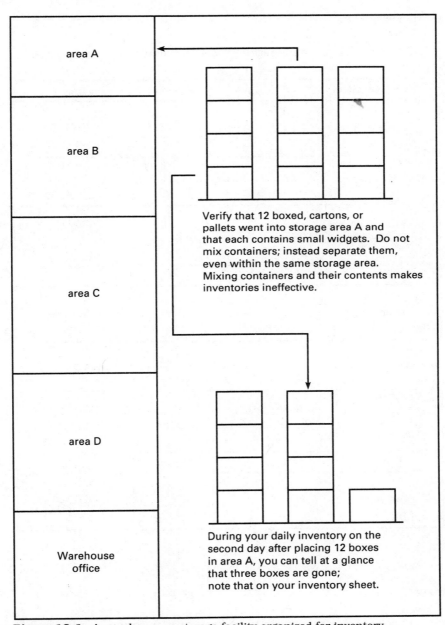

area A

area B

area C

area D

Warehouse
office

Verify that 12 boxed, cartons, or
pallets went into storage area A and
that each contains small widgets. Do not
mix containers; instead separate them,
even within the same storage area.
Mixing containers and their contents makes
inventories ineffective.

During your daily inventory on the
second day after placing 12 boxes
in area A, you can tell at a glance
that three boxes are gone;
note that on your inventory sheet.

Figure 12-1. A warehouse or storage facility organized for inventory.

Warehouse Inventory Worksheet			
Shipment/item	Area	Count	Date
large widgets	A	50	10/21/00

first day inventory

Warehouse Inventory Worksheet			
Shipment/item	Area	Count	Date
large widgets	A	50/45	10/22/00

second day inventory — note first count/then current count

Warehouse Inventory Worksheet			
Shipment/item	Area	Count	Date
Large widgets	A	50/30	10/23/00

third day inventory — down 20 boxes from first day

Figure 12-2. An inventory count sheet and procedure

percentage of all items) needs to be compared with the balance shown on the inventory sheets and records of employees. If there is a difference, make an investigation to determine why.

You should select your most trustworthy employees to conduct the inventories. Keep in mind, however, that even the best may not always remain honest unless they know that periodic unannounced verifications will take place. It's important that the supervisory count be conducted impromptu, following no specific pattern.

Remember that a supervisory count needs to include random physical inspections. It is not enough to count boxes, cartons, or pallets. Often, cargo thieves replace a stolen box with an empty or partly filled box, and reseal the box so that it appears to be full.

One astute warehouse thief watched the inspecting supervisor carefully. To verify that the cartons contained washing machines (a regular commodity at this facility), as they appeared to, the supervisor pushed them to make sure they did not budge. The next time around, the supervisor did the same, and satisfied that the cartons were not empty, moved on. The supervisor did not know that the warehouse thief had stolen some washers and nailed the empty cartons to the floor. In other cases, thieves replaced stolen items with scrap metal and bricks to equal the approximate weight of the contents stolen.

It is important to maintain your original inventory worksheets and records *outside* the warehouse. Your daily inventory supervisor might do a good, honest job. However, others who have access to the original inventory records (those records filled out in ink each day) can change them to cover cargo theft. This cannot happen if you keep the original records elsewhere. Any alteration of records will then have to be done with copies. Audits plus periodic supervisory inspections will detect copies quickly. When your policy is to trust employees who have access to cargo, but also to verify their work, you effectively eliminate the opportunity to steal.

Some costly business frauds involve nonexistent stocks of merchandise. One bank made a series of business loans on the imaginary contents of storage tanks. No one bothered to check. The message is clear: Simply turning on a spigot or looking at a gauge instead of assuming the contents can make a difference in protecting cargo losses.

Security Technique 132: Be Careful When Using Temporary Help

Some businesses hire temporary stockroom or warehouse personnel during high-volume periods. If you hire temporary employees, keep in mind that you may also be hiring built-in losses. Temporary employees know that they have no future with your business and that they will soon become unemployed again. They may easily take advantage of opportunities to pilfer cargo. Make sure that you place temporary employees in the least opportune positions. Whenever possible, have a regular employee work alongside each one.

The Ripple Effect of Cargo Losses

Business asset losses increase sharply when property and material become shipments in transit. The process of loading and unloading, the compartmentalizing of cargoes in ships, railroad cars, and aircraft, the movement of such carriers all present loss hazards of varying degrees. In each instance, you need to evaluate all contributory factors in order to create an effective loss prevention system.

If your business involves consigning your product or commodity to shippers, direct liability for loss may fall only to the shipper or its insurance company. However, you need to recognize the *ripple effect* of such losses. When cargo you consign (turn over to a carrier) does not reach its intended destination intact and on time, it will indirectly create losses to your business.

Suppose, for example, that one of your best customers orders 20,000 widgets from your firm. Your client probably has its own customers to satisfy, and it depends on your widgets as items for resale or for manufacturing material or components. When your widgets arrive and the count is 15,000 instead of 20,000 because someone stole 5000 widgets in transit, your customer will not be happy.

The ripple effect (indirect loss) on your client is a loss of business profits or a loss of customers. Chances are that your customer will begin looking for a new, more reliable supplier of widgets. If you lose the account, or if the customer begins dividing its orders among other widget suppliers, your business could head toward bankruptcy. Never underestimate the indirect losses created by pilferage and theft of your merchandise from consignees (carriers).

If your business involves warehousing, shipping, operating depots and terminals, using carriers, or insuring or otherwise underwriting intransit cargo shipments, the following information will be of help. You need solutions that reduce, control, and eliminate direct and indirect losses stemming from cargo diversion and pilferage.

Security Technique 133: Use the Inventory-in-Motion Concept

The inventory-in-motion concept is a key preventive measure for controlling cargo theft. According to this concept, it is "sitting" inventory or merchandise that breeds the greatest opportunity for theft. Thus products, materials, and other commodities should go as directly as possible from their source (your business) to the ultimate user (your customer), thereby avoiding storage depots, staging areas, and terminals. Although these facilities cannot be eliminated, your dependency on them for intransit cargo diminishes dramatically. Expanded application of the inventory-in-motion concept can save businesses of all types the rising costs of direct and indirect losses from cargo theft.

Security Technique 134: Determine If a Carrier Provides Adequate Services

The adequacy of a carrier's services depends on your priorities and the distinctiveness of your shipments or those consigned to you. You must evaluate the effect that cargo characteristics and other factors have on creating a lucrative target for theft. Then analyze the opportunity for diversion or pilferage that a particular carrier might inadvertently provide.

While on assignment as a criminal investigator in Japan, I was responsible for looking into reported losses of U.S. Department of Defense cargo entering and leaving Japan. I began to receive many reports of pilfered shipments that had been sent in sea-van containers. The combined losses totaled millions of dollars. The sea-van containers, owned and supplied by a variety of carriers, were certified by the U.S. Customs Service as "entry-proof" if they were fitted with locks and seals. Each of the shipments reported as pilfered was fitted with locks and seals at the point of origin and determined intact on arrival at its destination. The construction included reinforced heavy-duty steel, and the only access to the sea van's interior cargo compartment was through a set of double-interlocking doors. Thus everyone concluded that dishonest employees must have loaded the containers.

To create a test or control standard, I oversaw the loading of five containers with commodities bound for Defense Department retail facilities in Korea. I counted, inspected, and weighed each carton and affixed my initials to it. After each container was filled, I closed the container doors and placed two high-security Defense Department combination locks (locks with changeable combinations, used to secure classified information) and two randomly chosen serial-numbered metal seals on the doors. This excessive precaution assured me that since only I knew what combination would open the locks, and since the seals came from a different location, there was no possibility of tampering. The containers were later loaded onto the ship, and I went aboard to verify that no tampering took place.

Subsequently, I flew to Korea to witness the unloading of the ship. I again inspected the locks and seals and found they were intact, with no indication of tampering. However, upon opening the containers, I discovered that one-fourth to one-half of the contents were gone. This amounted to nearly a million-dollar loss. I sent the locks and seals to a crime lab for examination. The laboratory technicians agreed that the locks and seals had not been tampered with.

Later, my investigation revealed the source of the problem. Each sea-van door had three massive hinges fixed to the container with equally massive Phillips-head bolts. Using a large Phillips screwdriver and some muscle, thieves had unscrewed the bolts so that the entire door (with locks and seals in the center closing area) swung open on the hinges of the other door. After the theft (on board the ship), the door was swung shut and the bolts were replaced, leaving no trace of entry. When I reported this information to the Department of Defense, it created quite a stir. In Japan and around the world, crews of welders descended on the containers owned by the various companies, and spot-welded the bolts to prevent thieves from removing items. After this disclosure, the losses from sea-van containers ended.

I use this illustration to point out that if some of your merchandise disappears from seemingly secure containers or areas, do not jump to conclusions about the perpetrators or the location of the crime. You need to examine the situation carefully and conduct discreet control tests to learn exactly what the problem is, pinpoint where it lies, and then take corrective action. Be certain the opportunity to steal cargo, entrusted to you as a consignee or as an insurance underwriter, does not exist. As the shipper, you need to examine the modes and methods used, including the physical security measures needed to ensure that your cargo reaches its destination intact. These precautions will prevent those indirect business losses I described earlier.

Transportation Mode Evaluation Guide

Often your shipments, or those consigned, will move through more than one mode of transportation, such as from ship to train to truck. Consider the options that best eliminate the most number of movements and transfers. Evaluate the cost of transporting shipments using only one mode whenever possible. The cost of using a single mode may be higher, but if the cargo is diverted or lost by using multiple modes, there will be no savings.

Before choosing a transportation mode for your cargo shipments, consult the following guide.

Motor Transport

Motor carriers can serve as a primary or supplementary mode of cargo transportation. In a primary mode, motor transport is best for local or regional distribution operations. For long shipments, motor carriers can also be effective following inventory-in-motion concepts, especially when you own the vehicles or trailers and can ship a full load. Even if your cargo makes only a partial load and you share the trailer capacity with other shipments, motor carriers will usually supply a fast and reliable means of cargo transportation, with reasonable security.

Key Capabilities.   Motor carriers provide the most flexible mode of transportation over trafficable terrain in nearly all weather. They are also capable of transporting nearly any commodity with a variety of specialized equipment. Most carriers can handle distinctive cargo shipments (such as heavy equipment or steel beams) that do not conform to common trailers. When geographic location or terrain is not an issue, motor carriers supply the most economical and secure means of transportation.

Limitations and Disadvantages.   Route interferences and weather, terrain, and other obstacles in over-the-road operations may limit motor transport advantages, especially if the cargo is perishable or requires urgent delivery. If you are shipping extremely heavy commodities, motor transport may become uneconomical since these items will need several trucks, and even specialized vehicles. As costs escalate, weigh the advantages against other transportation modes.

Rail Transport

Shipping commodities by rail is a viable and often economical alternative to motor transport, especially for large, heavy, or bulky items. In manufacturing, a railhead located at the plant site offers secure loading and sealing of boxcars.

Key Capabilities.   Rail transport can handle a variety of commodities, and movement is rarely affected by weather. Depending on the point of origin and the destination, rail shipments can supply a reasonably fast, secure, and reliable means of transportation. Most services also have a variety of options—container, tractor-trailer, trailer, and private rail car—integrated into the rail system.

Limitations and Disadvantages.   Rail transport will lack feasibility for some cargo shipments because of geographic limits. For example, if your shipment must go to a destination off the main lines of rail transportation, the shipment might have to wait in marshaling yards for smaller rail lines to pick it up. Other disadvantages include the need to unload and store goods waiting for pickup by motor carriers, sometimes beyond the destination. This creates backhauls and delays, in addition to increasing the opportunity for cargo diversion or pilferage.

Water Transport

Surface or water transportation carriers include oceangoing ships, water craft on large lakes and along the Eastern and Western seaboards, and craft on navigable interior waterways (rivers). Each has a unique place for intransit cargo and can become either a primary or supplementary mode of transportation. Except for international cargoes that must use ships, localized or regionalized water transport may supply a good means of transportation and is especially suited to bulk, heavy, and outsized cargo.

Key Capabilities.   Water transport can be supplied under all weather conditions, except when the waterways are not navigable. It is feasible for any commodity and is highly economical for long-distance transport, especially when the route of the cargo is adjacent to a waterway.

Limitations and Disadvantages.   Water transportation is slow. Its flexibility is limited by the adequacy of terminals, waterways, facilities, and channels. The slowness and openness of this transport method make it vulnerable to cargo diversion and pilferage.

Air Transport

Air transport serves business with a primary, major supplementary, or complementary mode of shipment. It offers good security and rapid movement from origin to destination. Although more expensive than other modes, air transport can provide cost-effective inventory-in-motion benefits—namely, minimum handling and storage and rapid delivery from source to destination. All these serve to minimize cargo losses.

Key Capabilities.   In addition to speed of delivery, air transport has significant flexibility in that it is not hindered by terrain and surface obstacles. Considering the further advantage of security, air transport may be the most economical carrier for some commodities. If your shipment has limited shelf life or is ordered on a priority basis, air transport assures excellent speed and protection.

Limitations and Disadvantages.   Weather can affect the operational capabilities of air cargo movements. Also, destinations are limited by lack of air facilities in specific geographic areas or lack of cargo service into those areas. Other problems include weight and size restrictions and limited facilities for handling special shipments.

Finally, the cost of air transport may be too high. If the commodity has low priority or is of a size and weight that calls for committing one aircraft solely to the cargo, air transportation will not be cost-effective.

Security Technique 135: Minimize or Eliminate Rehandling of Cargo

If you ship or consign to ship cargo that has a high risk of pilferage because it's desirable for resale on the black market, you should choose the most direct routes and modes of transportation possible. When you evaluate your choices, reconsider the concept of "inventory in motion" and plan your cargo movement accordingly.

Security Technique 136: Ensure Effective Cargo Security

To provide security for intransit cargo, you must clearly delineate the responsibilities of carrier and consignee. Generally, the individual or firm in possession of the property is responsible for protecting the cargo in transit. However, responsibility varies with the size of the shipment and the mode of transportation. For small shipments of less than total capacity of the transport vehicle (truck, rail car, ship, or aircraft), the carrier's responsibility begins when its agent signs the shipping documents that acknowledge receipt of the goods at the shipping point. The responsibility ends when the shipment is delivered to another carrier or consignee.

For large shipments, the responsibility of the carrier depends on the type of transportation used. Not all carriers can be considered a secure means of transportation. For example, with international cargo shipments, the responsibility for cargo security transfers to foreign nationals whose interests and loyalties may not coincide with those in the country of origin. Some carriers provide little or no security and rely on the shipper to supply the needed protection for vulnerable cargo.

You need to ask, inspect, and evaluate these factors. Also, ensure that you have a full understanding of intransit cargo security responsibilities. Don't take anything for granted. Weaknesses may exist only because no one showed enough interest to stipulate that the cargo should have adequate security arrangements while in transit. Do not accept or consign any goods for shipment until all responsibilities are clearly delineated.

Security Technique 137: Guard
Against the Vulnerable Points
of Cargo Movement

During the movement of cargo, the loading and unloading points are the most vulnerable to theft. When a consigned carrier stays at a standstill, in the presence of laborers with handling equipment, opportunities for cargo diversion and pilferage reach a high. If you are consigning your shipments to a carrier, or if your business involves moving cargo or underwriting that movement, you need to ensure that loading and unloading is accomplished expeditiously with a minimum of activity. Review the policy and procedures of the responsible party carefully. Make sure that sound loss prevention techniques are applied to the loading and unloading of cargo.

Passive Techniques to
Prevent Cargo Loss

Protective management of business cargo in transit or in storage requires four important passive loss prevention techniques. A detailed discussion of each technique follows. Consider how to apply or adapt these general approaches to your operational situation.

Security Technique 138: Use
Shipment Identification
Markings

Distinctive markings serve to identify the cargo contents of containers, railroad cars, trucks, and other types of cargo transport. Sometimes, by regulation or other requirement, cargoes of certain characteristics must be divided into classes and marked with distinctive symbols. Flammable or toxic commodities are a good example. Often it is advantageous to code certain cargo (with numbers or other symbols) rather than identify the products directly.

Security Technique 139: Put
Locks and Seals on Cargo
Shipments

Any type of container that has a locking device needs to *remain* locked to protect the cargo and should have the added protection of a serial-numbered seal. Although a determined thief can force entry despite a lock and seal, the pilferage will be quickly and easily detected. For many would-be pilferers, the risk is just too high. Cargo locks should also have a serial number or be marked in such a way that a perpetrator cannot cut the lock and replace it with an identical one. However, serial numbers on locks and seals are effective only if they also appear on the bills of lading. If your business ships an entire load within a truck, container, or boxcar, you should place the locked containers on the shipping conveyance and mail the keys to the destination separately. Ask the recipient to notify you when the keys are received, and confirm the padlock and seal numbers on the conveyance on arrival.

Security Technique 140: Place Tarpaulin Covers over Cargo

Be sure that carriers protect your business assets against weather, diversion, and pilferage by using a tarpaulin or similar cover. The tarpaulin must be secured firmly to prevent easy access. To ensure that the tarpaulin stays in place, use nail-down wooden cleats and banded wires or wires crisscrossed from one side to the other along the top of the cargo load.

Security Technique 141: Containerize and Band Intransit Cargo

The best way to secure small boxes or packages is to consolidate them in a metal or wooden container (metal is best), such as a shipping van or sea van, described earlier. These receptacles will bear the name "container." When this approach is not feasible for a cargo shipment, "palletize" small items if possible and always band them.

When you mix categories of commodities on pallets, ensure that the most lucrative items for pilferage or the most valuable boxes form the "core" of your arrangement. When you load a container, truck trailer, or similar conveyance, place the most valuable and vulnerable items at the front or in the most secure area. Surround or follow these items with less lucrative and valuable cargo. This technique prevents easy access, discourages pilferage, and minimizes the opportunity to steal your most valuable assets. When practical, use coded descriptions (usually numbers) on boxes, invoices, and other shipping documents accompanying the cargo to prevent identification of the commodity by potential pilferers.

Quick Reference Guide to Chapter 12: Protecting Your Cargo

Several basic preventive techniques can reduce the opportunities for diversion or pilferage of business cargo. These techniques are the foundation blocks for building a cargo protective management program that satisfies all your needs. Look for weak or strong points in your operations, just as a thief would look for them, to determine if there are any opportunities to steal.

- Do you have a procedure for ensuring strict accountability of all cargo?
- Are facility records subject to periodic audits?
- Are sufficient and comprehensive physical inventories conducted?
- Are inventories conducted by activity personnel? Are they verified by senior management?
- What is your track record of loss, as shown by documented receiving, inventory, and shipping records?
- Are there weaknesses in your present system of physically unloading and storing cargo?
- Do employees make accurate tallies? Are the counts verified by a second employee?
- Are delivery records checked against shipping documents and purchase orders (where applicable)?
- Are shipping and receiving platforms free from trash? Are shipments neatly stacked for proper observation and counting?
- Are workers and visitors to the areas controlled properly?
- What passive and active physical security measures are in place in cargo areas?
- Are consignees or handlers of cargo required to produce photo identification and sign documents? Are signature cards on file?
- Have clearly delineated lines of responsibility been established?

13
The Fine Art of Hiring Employees

Most businesses do not screen their employees before hiring them. After that, unless a business is willing to face some devastating lawsuits, it's too late to fire them. Some businesses become defendants in court and regularly lose to applicants who know the law or find good lawyers who do. Hiring employees requires careful groundwork. You must screen applicants if you want to stay in business and make a profit. It is essential to screen applicants legally and, more important, fairly. The best general rule of applicant processing is this: Treat others as you would want them to treat you. If you follow this advice, your hiring practices will rarely be questioned, and if they are you will be able to defend them successfully.

Security Begins with the Hiring Process

In this chapter I will outline some dos and don'ts in hiring. Following these suggestions is the best way to select productive employees and at the same time to protect your business assets. Regardless of how meticulously you choose your employees, there's never a guarantee that they won't become thieves. However, when you fail to select employees carefully, the probability of theft increases.

In the next chapter, I will show you a secret weapon for protecting your assets through employee training. However, training will not supply the desired results unless you begin with suitable employees. All the countermeasures I presented in this book will not prevent dishonesty unless employees are selected properly to begin with and then trained effectively to do their jobs. Only then can you apply the preventive measures I suggest to eliminate debilitating losses from the myriad of criminal enterprises attacking businesses.

Security Technique 142: Know the Pitfalls of Using an Outside Screening Company

With a few hundred dollars, a computer, and a modem, anyone can set up a computerized preemployment screening company. A myriad of information brokers have gone into the on-line information business. Subscribers pay a fee to obtain an access code number and a telephone number to call the information provider by computer. Once on line, subscribers review a menu on the screen. After selecting the information they want, they complete a fill-in-the-blanks form requesting specific data. Within seconds after an applicant's name, social security number, and address are entered, information begins flowing.

As easy as this sounds, there's no guarantee of accuracy. Any information coming out of a computer can be only as good as that put into the computer. If the input operator makes a slight error, the information you get could be tainted or false. You may even get information about someone else under your applicant's name.

The good news is that businesses can (and should) do their own screening. Usually, screening applicants yourself will guarantee a legal and fair assessment. However, if you insist on using outside information sources, you will have to "screen" them as well to make sure they are skillful and reliable. If you hire information brokers who violate laws or regulations to get information for you and you use that information improperly (sometimes even if you do not use it), the courts will probably hold you jointly responsible. Also, obtaining information improperly could put you in a position of having federal or state criminal offenses filed against you. A stiff fine, jail, or both could follow this type of prosecution, depending on the severity of the misuse of information.

Security Technique 143: Screen the Screening Agency

This is what you need to look for in any outside agency that you hire to screen your applicants.

Check the state law where the "information broker" does business to see if that state requires licensing. Many states consider information brokerage as investigative activity and require a state-issued private investigator's license. Ask for the agency that issues private investigator and other licenses. If a state license is called for, inquire if the broker has one. If the broker does not, or if the license has expired, steer clear of the service. Also, ask the licensing agency if it has received any complaints about the broker or the broker's company, and request the name of the insurance company that has bonded the broker (most states requiring licensing also call for proof of a bond in a statutory amount). Call the bonding company and find out if it has a record of any claims or other adverse information against the broker.

After settling the licensing and complaint question, call the better business bureau where the screening business operates and ask if any complaints about the broker have come in. If there are complaints, request details.

Ask the person bidding for your applicant-screening business for five client references. Call the clients and get their opinion about the broker's work and reliability. Also ask how often the clients use the service and how recently they used it. Often, people will not criticize someone else over the telephone and will supply a good reference, although sometimes carefully veiled. If, for example, the reference tells you that it used the service several times but not in the last 3 years, this might be a way of voicing dissatisfaction with the broker. Bluntly ask the broker's references why they no longer use the service.

If the broker does business as a corporation or partnership, call the state corporation office (usually the applicable state's Secretary of State or Department of Commerce) and find out if the broker is registered and has current standing. Also, ask who incorporated the company and request names, addresses, and telephone numbers.

When you're satisfied with the firm, request a list of inquiries that it intends to make concerning your prospective employees. Ask about a specific fee. Also, ask for any paperwork you will need to get started. It is important to know how much the service will cost and exactly what you will get for your money. Does the fee include a credit history check? A competent information broker will ask you to sign a release form giving a valid reason for requesting a credit history on an applicant. If the service does not ask for such a form, the broker may be incompetent or operating outside the law. Steer clear of the service.

The best rule of thumb in this lengthy background investigation is this: If you uncover any derogatory facts about an information broker, find someone else. Better yet, as I've stated, do it yourself. Any business can obtain credit history information through local or regional credit bureaus, which also supply guidance on interpreting the information and on how you may legally use it.

Critical Steps in Preemployment Screening

Although it's not difficult to conduct a competent, legal, and fair background screening on an applicant, explaining the procedure properly to the applicant is another matter entirely. This is the fine art of hiring. Let's look at some preliminary steps.

Security Technique 144: Research an Applicant's Credit History

Obtaining a personal credit history from credit bureaus in the various areas where an applicant has lived is a solid tradition in preemployment screening. However, it is grossly overrated as important, regularly misused (sometimes illegally), and often misinterpreted. Misreading a person's credit history can, and regularly does, lead to false impressions. Remember that the credit history is designed to aid businesses or financial institutions in assessing a person's eligibility for credit. Extending that concept to include eligibility for employment creates a misuse (albeit legal) of credit information.

The Federal Fair Credit Reporting Act controls the use of credit information about individuals (businesses do not fall within the law) and stipulates how you can legally obtain and use a person's credit history. The federal statute allows use of credit history information when considering applicants for a job. However, some states have amended the federal law, making it more specific and placing added responsibilities on the employer (or its agent). For example, some states require that an applicant be notified of a credit history request and notified again when the report is received. Even when credit history information is legally obtained, you can get into civil or criminal trouble if you allow the information to fall into the hands of anyone not directly involved in the applicant assessment process. Other problems may arise if the information creates any type of discrimination—for example, if you obtain a credit history on one applicant and not on another or hire the person you did not obtain a credit history on. Here are some valuable guidelines for you to follow.

First, you must decide if the position you are filling "needs" a credit history. If you are hiring someone who will not have access to company money or easily pilfered assets, or someone who will not need keys to your business, it will be difficult for that person to steal anything of value from you. You do not need to conduct a credit history investigation on this type of application.

Even if the person you hire will have access to some of your business assets, you need to evaluate what that person could take and decide if the pressures of financial distress could make the employee steal.

If the person you hire will have plenty of access to your business money and other valuables, along with a key to the building, you will probably opt to have a credit history investigation. However, if the applicant is otherwise eligible, you should verify any derogatory credit information in the report directly with the creditor.

Security Technique 145: Safeguard An Applicant's Right to Privacy

Should you decide to obtain a credit history, make sure that the applicant knows your intentions and signs an acknowledgment. Also, verbally inform the candidate of your intent at the time of application and explain why a credit history is necessary. (If you do *not* intend to obtain a credit history, let the candidate know this as well, and make a note of it on the application.)

Remember that if derogatory information in the credit report becomes a deciding factor in denying employment, such information should first be verified with the creditor personally and the applicant should be directly informed. However, give the applicant an opportunity to supply his or her side of the story. Sometimes, a creditor will insist on a debt under unfair circumstances or when some type of litigation is in progress. Verify all sides of the situation.

Finally, retain an attorney to keep you informed about changes in employment law. Statutory law describes the particulars of a statute and remains unchanged except when officially amended. By contrast, case law (cases brought under or challenging a statute) evolves constantly as the courts interpret the spirit and application of the law. Thus you cannot rely solely on a statute as relieving you of any responsibility in hiring. Always check with an attorney competent in labor law before making any decisions (including obtaining credit histories), and before rejecting applicants on the basis of borderline or conflicting information.

Security Technique 146: Understand What "Bad Credit" Means

As noted above, credit history reports often become a deciding factor in preemployment screening, and many astute, loyal, and hardworking people are not hired because of "derogatory credit information."

Most businesspeople think that a person with "bad credit" is a bad risk and will probably steal business assets to keep the bill collectors at bay. By the same token, most businesspeople have high regard for a person with a good credit report because they perceive that the person honors his or her obligations and will become a reliable employee. I must stress that both conclusions are seriously wrong.

A person who has bad credit is not necessarily a "bad person." It is not difficult to have bad credit, especially in an era when status and success are too easily measured by how many credit cards a person has. When people lose their income for some reason, and stay unemployed for a time with no meaningful cash reserve to maintain their credit accounts, they must make choices. Food, shelter, and clothing head the list of items needing payment. If there is no other money left, the credit card bills will have to wait. If people default on several credit cards, and other bills go unpaid, their credit rating will be so low as to make them seem like the deadbeats of all time.

Defaulting does not mean that a person has no intention of paying his or her bills. Look for actions to determine a person's intent. For example, did the defaulter correspond with the creditor, explaining his or her financial situation and intent to honor the obligation in partial payments? (Sometimes, of course, creditors will not accept partial payments.) You need to know the person's intentions, as demonstrated by positive action taken to work out the situation.

Now let's look at the employment candidate who has a squeaky clean credit history but has been out of a job for months. Did the candidate manage to keep afloat by borrowing money, using a prudently saved cash reserve, or selling something valuable? Or is the applicant a dishonest person who steals, commits armed robbery, writes bad checks, or sells drugs to maintain an impressive credit record? My point here is that the report in itself does not tell you anything. You must carefully analyze the circumstances and merge your findings with other information.

Security Technique 147:
Interpret Credit Reports
Accurately

As you review an applicant's credit report, check each period of high credit with dates given in the applicant's employment history. The credit report will list the high periods of credit and not whether the payments arrived punctually. It will also give the current balance (usually about 1 or 2 months behind the current date) and the names and addresses (sometimes telephone numbers) of the creditors. You need to compare the amount of credit with periods of unemployment and then total the amount of payments called for (the total debt in the beginning and now). Next, weigh your findings against how long the person has been unemployed, if that has validity. (The applicant may now be working elsewhere and seeking a change of position.)

Suppose, for example, that the applicant is now or was recently employed at an after tax salary of abut $500 a week ($2000 a month). If the applicant's combined payments on credit amount to half or more of that, and the credit report shows no home mortgage, you need to exercise caution. Considering that the essentials of food, shelter, and clothing usually take more than half a person's salary, you may wonder how the candidate managed to maintain credit payments even while employed. However, do not jump to conclusions. Maybe the applicant has other income not shown on the application, such as a small business venture or a part-time job. Perhaps a working spouse contributes to the payments. Verify these other sources before assuming that the applicant is not qualified or responsible.

Now suppose your applicant has "bad credit." You need to do the same analysis to determine the reason for it. Maybe the candidate lives beyond his or her means, or just becomes carried away by the ease of credit. Such a candidate has worked for years and could easily pay his or her bills but chooses not to do so. The obvious conclusion here is that the person could be irresponsible and unreliable on the job as well.

However, what if the applicant has been unemployed for a time through no fault of his or her own? Perhaps a health or financial crisis took a large sum of money. The only solution for the applicant is to use the credit available to meet the need, and to repay it when he or she is again employed. Such a candidate may well be highly responsible and reliable.

If you rely on the credit history report to make your decision, analyze it carefully and ensure that you verify any problem areas. Again, credit bureaus do make errors. Whenever you come across seriously derogatory information, call the creditors directly, verify that the information is correct, and determine what actions they have taken to collect the money and what their future collection plans include.

Processing the Employment Application

Processing an application for employment begins with a solid, unambiguous application form. In addition, you will need to set detailed employment standards and a job description for each position you offer. Application forms and résumés should be verified, with the scope of verification increasing according to the importance of the position to be filled. Finally, performance tests and other evaluation aids, used judiciously, can help you find the best person for the job.

Security Technique 148: Develop a Good Application Form

Many job application forms seem to be designed by paper conservationists. They call for detailed information and leave spaces that cannot possibly accommodate that information. Applications for positions within your business should be designed to satisfy *your* needs, not those of a forms designer or recruitment specialist.

When you have a variety of positions in your business, consider tailoring your job application form to each. Also consider how in-depth your investigation of an applicant for each position needs to be. A candidate filling out an application form tailored to a specific position can provide you with precise information that will help your decision-making process. If you use a generic application, many applicants feel compelled to fill in all the blanks, and they may make inflated claims that taint your perceptions.

For instance, if your business position does not require a college education but your application calls for a listing of all colleges attended and all degrees earned, the applicant might claim to have a degree from a reputable college, fully knowing that you will not check. Putting that qualification on your application form serves only to sway your judgment unnecessarily. A college degree (fake or real) might be enough to make you choose that person over another.

Most office supply stores have generic applications that conform to the laws regarding what you can legally ask an applicant. You can use them as models for your tailored applications, but *do not add questions* and delete only those that do not apply. Again, ask a lawyer knowledgeable in the labor relations field to look at your tailored applications and make sure that you do not violate any law or regulation.

Security Technique 149: Set Employment Standards

Whatever the size of your business, you should have a detailed list of duties and responsibilities for each position. This is simply good business management. Both existing and potential employees need a clearly written set of employer standards, called a *job description*. This description not only makes it clear what job an applicant is specifically applying for but also defines exactly what he or she is supposed to do after being hired.

In your job description, be sure to avoid language or standards that give any hint of discrimination. For example, some businesses still require advanced degrees or diplomas for positions that could easily be filled by people with only a high school education. Unless the position absolutely calls for specific skills, education, or other relevant qualifications, do not include them in your job description.

Similarly, do not inflate the job in your magazine and newspaper advertisements. Many job advertisements, especially those in the Sunday paper displays, state that applicants need various types of degrees and significant experience. Yet the salary offered is at entry-level rates not much beyond minimum wage. If you need an experienced person, and one with considerable education, then set those standards, making sure that they are valid. If you inflate your standards beyond those needed and beyond the salary you are willing to pay, you will get applicants who inflate their qualifications accordingly, in the hope that you will not bother to verify them. Many businesses complain about inflated résumés or application claims. Information brokers · use these deceptive claims as a key advertising point, justifying intense background investigations. However, if businesses start with clear, honest, and fair job descriptions commensurate with the experience of the person needed in their operation, applicants may not exaggerate their qualifications.

Security Technique 150: Verify Applications and Résumés

Businesses can err two ways in verifying applications. Either they check too much and pay a considerable fee to information brokers or they do not bother to validate applicant claims at all. The depth of applicant validation should be commensurate with the job your business offers. Do not accept the premise that a formatted, standardized background investigation serves all. I always recommend that you begin the application process with a cursory introduction to the applicant. (I hesitate to use the often misinterpreted term "interview.") Giving applicants an honest and detailed job description sets the process on the right track. Applicants need to know exactly what you will expect from them if they become your employees, exactly how much they will earn, and what their chances of advancement are. An advertisement in a newspaper doesn't provide this three-dimensional view of the job. When you shake hands with job applicants, extend a copy of the job description, and talk briefly with them, they are likely to tailor their responses to the position they seek. This makes your verification task easier. Usually, applicants will also avoid inflating their qualifications to impress you.

As you receive applications for the position, screen out those that clearly do not match the qualifications listed on the job description given to each applicant. Once you have a handful of applications that match your qualifications, verify only those parts that have a relationship to the job description. (Again, leave credit histories as the last part of your verification process and make them only when it is absolutely necessary. Doing so will ensure that no bias perceptions creep into a process that should stay impartial.)

Your first step should be to verify the "character" of the applicant. Much can be determined by contacting past employers. If the person has just finished high school or college and never had a full-time position, usually he or she has filled part-time jobs. You should call former employers and schools. You need to determine if the applicant did work there, or was in school as claimed, and if so what the reference perceives the character of the applicant to be.

You need to use caution in this phase. For example, two references may make marginal or even derogatory comments about the applicant, while five supply good recommendations. Look at where the two unfavorable perceptions came from and consider if they were impartial or if there might have been some reason for the bias. You also need to analyze the favorable statements, and then weigh them against your added findings.

I once conducted a background investigation on an applicant for a government job. One reference (an elderly lady) had lived next door to the applicant for several years. She had a low opinion of the applicant and when I asked why, she said that he used to mow his lawn almost completely naked. Later, from interviews of other neighbors, I learned that the applicant had worked in an office during the week and on Saturdays mowed his lawn wearing a bathing suit to get a suntan. Always verify and clarify any derogatory information you receive about applicants.

If education is important to the position, ask the applicant to sign a release form addressed to the school, college, or university so that you can receive a transcript of his or her educational achievements.

Tests to Screen Applicants

I have looked at a variety of application processes used by businesses. Some give applicants competency tests, and others have applicants answer a variety of questions about their personal aspirations as related to the job. All these techniques, I have found, are basically useless. They tell you nothing of substance, and might sway your decision in the wrong direction.

The purpose of testing is to determine what the applicant is really like as opposed to the image he or she projects. Personality and other profile tests have a hidden pattern that savvy applicants can quickly identify, especially if they have taken such tests before. The secret is to answer consistently, even though the questions may appear to be different at first glance. Throughout the test, which is often several pages long, five or so themes persist and are related to such categories as sports, hobbies, honesty, loyalty, and stability. Supposedly, the employer or test administrator can evaluate candidates accurately from their consistency in answering questions that have the same theme. When a candidate does not answer consistently, the employer perceives that the person may be unstable or dishonest, or may not have motivation or aspirations for success. Like the old polygraph test, now outlawed for employment screening under federal law, these written tests often prove unreliable.

Other types of tests, with equally little value, ask the applicant to write short answers to a variety of questions: "Why do you think we should hire you?" "Why do you want this job?" Other questions focus on personal aspirations and motivations. An applicant will rarely answer such questions with candor, fearing that he or she will be excluded from consideration. For example, if an applicant says he or she wants the job because "I need the money," chances are the application will go directly to the files. However, if the applicant wants the job because he or she has "long admired the business as setting a standard of excellence and has wanted to be a part of the business's family of employees," the response will flatter the egos of those who make the decisions. If all the answers are slanted to the business and motivation is geared to working up "through the business," the applicant will be perceived as loyal and will have the job wrapped up.

The best test always comes from performance. Anyone can learn to take tests well. However, you want to know if an employee can perform the job with competency. If you must give tests to applicants, have them perform as best they can in the exact job they will later do if hired. The performance test can be a realistic mock-up. For example, if you want a cashier for a checkout counter, create a performance test using a basket of goods like those a customer would select and have the person check out the items. If you give a written test rather than a performance test, the applicant may appear qualified as a checkout cashier. However, you may learn later that his or her dexterity for operating a cash register fits the adage "all thumbs." You probably will have to begin looking for another employee. Finally, performance tests coupled with honest applications may uncover candidates who cannot do the job applied for but who demonstrate strong aptitude for another position.

The Pros and Cons of Applicant Interviewing

Like standardized testing, credit investigations, and background checks, traditional interviewing is fraught with pitfalls. You need to be aware of these traps and of ways to overcome them.

Like traditional testing, traditional interviewing often involves pointless questions: "Why do you want this job?" and the old favorite "Why do you think I should hire you?" These questions cannot be answered correctly. They indicate only that the interviewer has no idea what type of employee the business wants, and they place interviewees in the position of telling the interviewer what he or she is looking for. With this line of questioning, a skilled job applicant gets control of the interview and duly impresses the interviewer with answers manipulated to please. For example, when asked why he or she should be hired, the skilled applicant will go into a long speech about admiring the business and considering it an honor and a privilege to learn from the infallible boss (or bosses). Since one of the bosses is conducting the interview, the applicant has puffed the answer to swell the interviewer's ego. The traditional interview cannot inspire honesty. It is a poor way to begin with an employee who has rightfully perceived the entire process as borderline or outright dishonest. This approach can open the door to dishonest employees who will steal your business assets systematically or whenever the opportunity presents itself.

What, then, is a "useful" interview? It is one that focuses on what you need from an employee in the position applied for. It is one that gets the applicant to discuss how excellent performance might happen if he or she were hired.

Suppose, for example, that you are trying to fill the position of checkout cashier. Your first task is to assess the characteristics and qualifications of the applicant. Next, lead into a discussion of how your business presently handles customer flow and customer relations. This will put the applicant at ease for a key question: Ask, based on the applicant's experience as a cashier and as a customer, how he or she views possibilities of improving the job or customer relations. If you conduct the interview with this type of meaningful discussion and questioning, you will create respect and rapport while having an in-depth look at who the applicant really is.

If the applicant appears to be tuned to your philosophy of business generally and seems amenable to learn and conform but is not afraid to offer suggestions, and if all other information is favorable, you may have found a new employee. Conversely, if the applicant decides that he or she does not agree with your business philosophy, the applicant may say as much after a friendly, professional conversation and go off to seek other employment.

In this last-stage interview, you are trying to ensure compatibility among your business operations, existing employees, and the applicant. Generally, when you use the discussion approach, the applicant selected will be a good employee, and those rejected will not perceive unfair treatment or discrimination. Instead, they will conclude that you gave them every opportunity to express themselves honestly.

Security Technique 151: Select Employees for Reasons Beyond Personality and Skill_____

The last step in your applicant selection process is to determine how well a potential employee will fit into the total business operation. All business success stems from teamwork. The greatest business leader cannot singlehandedly ensure a profit. Instead, the employer relies on employees to follow his or her lead and perform productively.

You can determine if a candidate has teamwork traits in several ways. First, ask questions pertaining to teamwork when verifying past employment references. Whenever possible, get the names of the applicant's former supervisors and managers and talk to them about the applicant's character. Then inquire how well the person fit into the work environment as a team player. Some people have much to offer but cannot work productively with others.

Also look for "teamwork tendencies" during your first contact with the applicant and later during the last-stage interview. For example, you might move the interview discussion to the concept of teamwork and ask the candidate's opinions about former places of employment and the other employees who worked there. People categorized as "loners" in the workplace have a variety of telltale characteristics, although few recognize these traits in themselves. You need to search out candidates who:

- Are better "talkers and tellers" than listeners
- Have a tendency to dominate conversation
- Try to overwhelm you with personal achievements
- Are overly qualified for the position
- Display strong leadership traits, with a strong leadership background
- Have a strong ego and will, as demonstrated by conversation

These and other traits do not in themselves make a candidate peculiar or unsuitable. However, they do suggest that an applicant may not fit in as a team player. Your concern is to hire in the best interest of your business. An applicant with otherwise impeccable credentials may not meet standards that you know have importance to your operation. When a person demonstrates all the traits noted above during an interview, background checks will probably confirm that he or she will make a great manager but not a good worker. Even as a manager or supervisor, the candidate would probably move in his or her own direction instead of yours. Most people who are "overqualified" and have strong leadership traits generally have the best success in carving out their own business or operating alone in some craft or profession that allows them to work where they excel.

Quick Reference Guide to Chapter 13: Screening Job Applicants

- Use caution in contracting with applicant-screening companies. Screen *them* before allowing them to represent you in checking the backgrounds of applicants.

- Whenever possible, do your own employee screening.

- Do not depend on an applicant's credit history, because it is only a part of the total picture.

- Remember the rights of applicants and be careful what you ask of them.

- Create meaningful job descriptions that set viable standards.

- Verify the information that a job candidate supplies on the application form.

- Create a set of application forms, tailoring each to a specific job description.

- If you must test the applicant, choose those tests that make sense and directly relate to the job.

- Conduct conversational or discussion interviews before offering anyone a job.

- Determine if the applicant will operate as team player.

14

Training Employees: The Secret Weapon of Asset Protection

According to recent studies, U.S. companies with 100 or more employees spend nearly $50 billion each year for direct training. Certainly, training employees to perform their jobs with greater expertise and productivity is valuable. However, rarely in this massive training effort are any programs devoted to protecting an employer's business assets. Smaller businesses, which have few if any training programs, need to rethink their values and focus more of their time on creating a secret weapon to protect their assets through training programs.

In this chapter, I will show you how to bring an employee training program to life, making it an intrinsic part of your asset protection strategy. As you do so, you will increase efficiency and productivity, and enlist your employees in helping you earn and keep your business profits. The training techniques focus on the protection of business assets. However, you may apply the same methods to any business area, including sales, productivity, and customer relations. You do not need to create an added expense by hiring professional trainers, although they are helpful. You can train employees yourself.

If your business organization has executives, managers, supervisors, and senior employees, the techniques described below will help them become effective trainers. This unique employee training program can work for individual employees or employee teams and adapts to any business environment.

The Critical Importance of Employee Training

You need to ask yourself an important question about why your employees do not perform well and why they become involved in pilferage or turn a blind eye when other employees, vendors, or patrons steal from your business. Creating effective deterrents and countermeasures will eliminate most losses. However, you need the final secret weapon, the last ingredient that creates attentive employees: a security-oriented training program. Contrary to the beliefs of many business owners, employees do not know inherently what they should do and how they should protect business assets. You must train them and then create performance standards that define the parameters of their jobs.

Security Technique 152: Use Training to Solve Business Problems

When you develop training objectives and institute employee training programs, especially ones related to job performance and business asset security, the immediate advantages become obvious as profits increase and losses dwindle. Other advantages and benefits will also emerge. When you choose your employees properly (see Chap. 13), you create a foundation on which to build. A sound training program gives employees a sense of achievement and satisfaction that helps them take a genuine interest in the business beyond protective measures. Employees also build confidence and gain a clear understanding of their jobs, why their work is important, and how they contribute to the success of the business venture. Finally, integrating asset protection into your training program will help employees recognize their responsibility to surpass the mechanics of the job alone and work to ensure that the company thrives instead of being riddled with theft and other crime.

At the core of professionalism is expertise in a given job, craft, or endeavor—a set of skills acquired through an employee's personal efforts and through training provided by an employer. However, personal efforts and training can build expertise only when employees work toward standards of job performance that they understand. The key to developing professional attitudes and performance is to delineate clearly to employees that they need to meet *your* business standards of performance.

Security Technique 153: Build Your Training from the Basics

A successful business always emerges when employees are fully trained in the fundamentals of business operations. Whatever training you conduct, never forget that the ability to build professionalism, confidence, motivation to excel, and interest in protecting business assets depends largely on how well employees execute the fundamentals of their jobs.

A Performance-Oriented Approach to Training

Traditional employee training methods rely on lectures, conferences, seminars, and demonstrations as primary tools. The proven methods and techniques supplied in this chapter, however, focus on performance—enabling you to provide training more efficiently and effectively. Here is why you need to install this performance-driven and security-oriented program without delay.

The greatest problem with lectures, seminars, and other conventional training techniques is that they focus the employee (student) on the instructor. A second problem of conventional training is that employees are bombarded with information that exceeds their retention level. Most employees in training will remember only 10 percent of the information presented, and often it is unclear to them how they can apply event that 10 percent to their jobs. A third problem is lack of dynamic course content. Finally, conventional training does little to give employees the practice and feedback they need to acquire new skills.

Differences Between Performance-Oriented and Conventional Training

- Conventional training relies on the lecture as its chief method of instruction. Performance training uses short demonstrations and "learning by doing" as its main means of instruction.

- Conventional training places the instructor in the central active role. Performance training makes employees active and centers training around them, giving them the time and support needed to learn. The trainer acts as catalyst and supervisor.

- Conventional training selects content according to what the instructor can present or cover in a certain amount of time. Performance training digests content into a set of high-priority skills important for employees to learn in the time allotted for training.

- Conventional training uses "grades" to rate what the employee has learned. Performance training sets standards all employees must meet. When they are unable to meet the standards, employees practice until they can.

Advantages of the Performance-Oriented Approach

Four key elements contribute to the unparalleled effectiveness of performance-oriented training. You need to follow and apply them as you consider creating or renewing your business training program.

Precise Training Objectives.   Performance-oriented training is characterized by objectives that state precisely what the employee must be able to do upon completion of the training. A properly structured training objective has three elements:

- The *task* to be accomplished
- The *conditions* under which the task is to be accomplished
- The *training standards* of acceptable performance

With a precise expression of task, conditions, and standards, employees understand that if they can already perform the task to acceptable standards, that is good; if not, they know what they must learn to achieve the standard. Developing these precise objectives can also benefit you. You will understand what the training entails yourself.

Job-Related Training Objectives.   Because training is preparation for job performance, your training objectives must relate to the job requirement of the employees being trained. For example, a cashier and a stockroom clerk both work for the same business; however, each has a very different set of responsibilities. Although some aspects of your training might include all employees, separate training is necessary if it is to be effective.

Training in an Active Business Environment.   In performance-oriented training, lectures and conferences are minimal. Most training time is used for controlled practice of a task (learning by doing). Such practice ensures that employees in training become capable of effectively performing the tasks called for by their jobs. Because the bulk of training time is devoted to employee practice, the standards can be better met. Further, because the training objectives relate to the specific job of an employee, he or she will be better able to understand the need for training.

Continuous Feedback on Progress.   When you place emphasis on "learning by doing," you receive continuous feedback on the progress of those in training and on the efficiency of your training program. There is no need to wait for the end of training to test results. (The end of a training session is a bad time to find that little or no learning has occurred.) Further, because employees play an active role in training, they receive a steady flow of indicators on how well they are progressing. And because each objective has a training standard, the employee can receive immediate feedback and reinforcement that he or she has accomplished the objective. This feedback enables the employee to progress from simple to difficult tasks.

Security Technique 154: Develop a Performance Orientation

Performance-oriented employee training has an overriding objective that can best be stated in question form:

"Does this training really prepare employees to do their jobs?"

Or, to focus on security or business asset protection, the question can be slightly modified:

"Does this training really enable employees to better protect business assets from (name a specific type of crime such as shoplifting, armed robbery, or pilferage)?"

The performance-oriented approach to employee training, as presented throughout this chapter, forces clear and precise thinking about training preparation for job performance. An added benefit is that anyone can serve in the role of trainer. Notice that I used the word "trainer," not "instructor." This is an important element of the performance-oriented approach.

Here are three more questions to ask yourself as you build a performance orientation into your training program:

"Where am I going? What must my employees do as a result of their training?" (*desired results of training*)

"Where am I now? What can my employees do now compared with what I want them to do as a result of training?" (*current versus desired level of training*)

"How can I best get from where I am to where I should be? What techniques, training methods, and organization offer the most effective and efficient use of available resources?" (*how training is to be conducted*)

Security Technique 155: Understand the Importance of Learning and Motivation

Your training must increase an employee's knowledge, skill, and performance. Five conditions are a prerequisite to learning. Employees must:

1. Realize they need training
2. Understand what you expect them to learn
3. Have an opportunity to practice what they have learned
4. Get reinforcement that they are learning
5. Progress through training in a logical sequence

If these five conditions are met, you have created an environment conducive to learning. Beyond that, of course, employees must be ready and willing to learn. The characteristics of performance-oriented training help to establish optional learning conditions.

Security Technique 156: Set Clear and Explicit Training Objectives

Generally, performance-oriented training begins and ends with your training objective. For a given skill or application, a properly structured training objective can also contribute to the evaluation process. To describe and remember this performance approach, use the following equation:

$$\text{Training objective} = \text{Training} + \text{Test} + \text{Evaluation}$$

A complete training objective answers three important questions:

1. What skill do you want employees in training to acquire? (*task*)
2. Under what circumstances do you wish employees to demonstrate the skill they acquire? (*conditions*)
3. How well do you expect employees to perform? (*training standards*)

With precise training objectives, you and your employees have an understandable, common focus for the training effort. This reduces to a minimum the varied, occasionally conflicting interpretations that arise from using such terms as "proficiency," "familiarity," and "working knowledge" to describe the desired results of training.

Security Technique 157: Use Training Resources Efficiently

Because your training objectives appear in precise, measurable terms, you and the employees you train develop an understanding of exactly what the performance-oriented training will achieve. The training resources (employees, money, and time) necessary to complete each objective will become more accurately estimated and planned. This factor becomes increasingly important as you prepare your training sessions.

Security Technique 158: Focus on Job-Related Tasks

Throughout performance-oriented training, employees perform job-related tasks under specified conditions until they achieve the level of proficiency established by your training standards. Those employees who quickly master a particular training objective may shift to help slower learners (a form of peer instruction), or they may continue their own skill development in other areas. The training remains focused on employee performance instead of on the trainer's ability to present instruction.

How to Create a Business Training Session

The following techniques are essential to ensure that you properly prepare and conduct performance-oriented training. When you make them part of your training program, the process will become both effective and efficient.

Security Technique 159: Describe the Desired Results of Training

What is your starting point when preparing your business and asset protection training? As noted earlier, you need to create one or more specific training objectives that develop the mechanics of task, conditions, and standards. You also need to determine to whom you will give training and the reasons your employees need the training. Decide on the date, time, and place of the training sessions.

Security Technique 160: Set Intermediate Training Objectives

Despite the time and resources available to you, you cannot provide employees with total training in one session. Your training effort must have systematic increments, created through intermediate training objectives. For example, if you intend to train your employees in techniques of eliminating shoplifting or in ways of protecting business assets in the shipping and receiving, you will need several training sessions. Your final training objective will be accomplished by the performance you hope to receive from your employee after training. You need to determine how to get from A (the beginning of training) to D (the end of training). Each session needs an intermediate training objective. After combining them all, you will reach the final training objective.

You can establish your intermediate objectives with the following steps:

A. Develop tasks needed to complete the training objective.

B. Establish the conditions under which each task must be accomplished.

C. Establish a training standard of performance for each task.

You must go further when preparing to conduct your training.

A. Determine which intermediate training objectives the employees cannot successfully perform without further training.

B. Organize the intermediate training objectives into a progressive sequence (simple to complex), consistent with your available resources.

C. Estimate the training resources needed to accomplish each objective.

D. Complete your administrative requirements, such as obtaining equipment, writing a lesson plan, scheduling, and developing facilities.

Security Technique 161: Conduct Training to Standards

Your final task involves conducting and monitoring the training to ensure that employees can perform your objective. The work you have done up to this point will culminate in benefits if you remember these important guidelines.

1. *Be an effective business-oriented leader.*   If you want your employees to take part enthusiastically in the training sessions, you must personally set an example. Your knowledge, bearing, appearance, manner, and desire to help are extremely important in creating a training environment that will make your employees "want" to learn.

2. *Ensure that employees meet established standards.*   Remember that employee training is the key to productivity and professionalism. Your effort to establish realistic, attainable standards for employees means that you must stay involved in the training, supervision, and evaluation process. Consider using the faster learners to help the slower learners.

3. *Confirm that employees can execute the fundamentals.*   If your employees cannot perform each of your intermediate objectives and satisfy your established standards, do not "press on" with the training. Only with a sound understanding of the fundamentals can employees go on to satisfy the total training objective. When this happens, they will excel in performing their job-related responsibilities and tasks, including protecting your business assets.

Security Technique 162:
Allocate Time and Resources
to Training

Each performance-oriented training session should be divided into three phases. Time and resources must be allocated separately to each phase.

Phase 1: *Explanation.*   You or a trainer states the training objective (or intermediate training objective) and explains or shows how to perform the objective.

Factors for consideration

- Current proficiency of employees: influences the time needed for your explanation and demonstration
- Number and complexity of objectives: influences the time needed for explanation and demonstration
- Number of employees to be trained versus resources available: includes time, facilities, equipment, devices, assistant trainers, and training aids such as videos and films

Phase 2: *Practice.*   Employees "practice" the objectives to acquire the proficiency called for by your training standards (keyed to on-the-job performance standards).

Factors for consideration

- Current proficiency of employees: influences the time they will need to learn or develop the skill. Other factors, that influence how quickly employees learn include motivation, intelligence, coordination, and morale.

- Number and complexity of objectives: influence the time necessary for employees to acquire proficiency. How quickly employees acquire proficiency also depends on their current proficiency.

- Number of employees to be trained versus resources available. How many employees can practice simultaneously depends on the available resources compared with the number of employees (e.g., one aid or device per person to practice with versus one aid or device per 10 people).

Phase 3: *Testing.* Without coaching or assistance from peers, employees perform the job-related training objective (or intermediate objective) to meet your established standards, which are keyed to on-the-job performance standards.

Factors for consideration

- Current proficiency of employees
- Complexity of training objective
- Number of employees to be trained versus resources available

Security Technique 163: Train Employees Collectively as a Team

Collective performance-oriented training prepares employees to perform team or group tasks essential to your business operations. In collective (teamwork) training, as with individual training, your objective remains the key to efficient and effective performance. Collective training objectives specify:

- The performance of an employee group task
- The conditions necessary for performing the task
- The team or group training standard that you desire employees to meet

Only by having precise training objectives can you and those in training clearly understand what you expect them to accomplish. To illustrate this important point, consider how a successful coach trains a football team. Before preseason practice begins, the coach and coaching staff carefully prepare the offensive and defensive plays that will enable the team to accomplish its purpose (to win football games). These plays create collective training tasks.

For example, to complete the training objective for a pass play, the coach sets conditions and a training standard for the team collectively. Specifically, the conditions might include the rules of football (size of the field, number of players, infractions, and so on), the yardage needed for a first down, the offensive team's field position, and the alignment of the defensive players. The team training standard is a completed pass that gains enough yardage for a first down. This standard specifies what the entire team must achieve.

Creating your collective business-oriented training, including asset protection as an integral part of your operations, calls for the same attentive effort.

Security Technique 164: Use a Multiechelon Approach to Collective Training

To ensure that your collective training sessions use available time and resources efficiently, adopt a multiechelon approach—that is, simultaneously train different elements of a group before putting it all together in the workplace. This approach enables you to train your entire staff regardless of their positions. The concept of teamwork during business operations becomes your training goal, while the objective remains developing job proficiency for each position. For example, you can train assistant managers and supervisors to learn and practice their skills without wasting the time of other employees. Meanwhile, you can have other employees learn how to perform their jobs under the direction of an assistant.

It is important to train each echelon of your employees individually until they are proficient. Then have them train to work together as they would function in day-to-day business operations. Consider this type of training for fine-tuning your entire staff in a controlled training environment using a performance-oriented approach so that errors do not cost valuable business or customers. You can also use the training sessions to identify the weaknesses and strengths of employees who need development. Often, employees may seem to know and do their jobs properly; however, during the performance-oriented training sessions, you may discover that their proficiency is seriously lacking.

Security Technique 165: Monitor the Progress of Employees

Your employees in training and on the job need to know the answer to the age-old question: "How am I doing?" With a performance-oriented approach to training, monitoring progress is simple. Because employees know what they must do, they have an established benchmark against which to measure their performance throughout the employee training sessions.

You too must participate by overseeing or supervising the training process. Monitoring training helps determine employee performance of intermediate objectives. Again, employees must properly execute the fundamentals and meet the established standards before they can perform the total objective.

It is important to distinguish between training effectiveness and training efficiency. *Training effectiveness* deals with how well the employees perform the objective. *Training efficiency* deals with how well you use your available training resources.

When your employees can meet the established standards, the training is successful. If employees do not meet your established standards, you must try to pinpoint the problems. Establishing clear evaluation techniques and finding reasons for problems within your training effort are difficult tasks. The following section will help you troubleshoot problems systematically.

How to Evaluate Your Employee Training Program

In performance-oriented training, you can focus on whether your employees can perform your objectives and meet, or exceed, your training and job performance standards. Whether your training sessions are successful or problematic, self-evaluation should remain a routine matter so that you improve the quality of training, bring it into better focus, or find out why it's not successful. Below is a checklist that is designed to help you assess this aspect of your employee training program.

Preparation for Training

1. Did you develop specific training objectives (including intermediate and total) and state them as task, conditions, and measurable training standards?

2. Did your lesson plan contain these minimum elements of information?
   - Your total training objective (the desired result of training)
   - All intermediate training objectives (if any) presented in logical sequence
3. Did you consider the following administrative details carefully?
   - When to conduct your training
   - Training location (acceptable, quiet, conducive to learning)
   - Whom you will train
   - Who will have responsibility for the training
   - Training aids and equipment to use
   - Key references (when applicable)
   - Training activity sequence and estimated time

Conduct of Training

Phase 1: Explanation

1. Did you:
   - Tell employees that the training objectives include performance standards?
   - Give employees clear reasons for learning and becoming proficient in the skill?
   - Show how to perform the objective from the employee's viewpoint?
   - Give your demonstration in a location where all employees could see clearly?
   - Demonstrate each step of your objective in the order to be performed?
   - Give all information necessary for performance of each step?
   - Where appropriate, call for employees to perform each step immediately after your demonstration and explanation?
   - Stress critical (key) points?
   - Avoid giving employees needless or unrelated information?
   - Pace demonstrations according to the employees' learning ability?

Phase 2: Practice

2. Did you:
   - Correct employees if they made errors?
   - Tell employees what to do when they needed help?
   - Show employees what to do when they needed help?
   - Require employees to perform all the steps or parts of the steps you demonstrated?
   - Prompt employees when necessary by asking questions such as "How do you do this function?" and "What must you do now?"

- Ask employees questions that help them understand critical points, such as "Why do you do that?" and "What would happen if?"
- If task results vary with conditions, give employees practice situations that differ from demonstration and walk-through situations?

Phase 3: Testing

3. Did you:
   - Explain performance testing instructions clearly and slowly?
   - Observe complete performance of employees being tested?
   - Avoid correcting errors before the performance test ended?
   - Arrange testing conditions so employees could not copy from one another?
   - Explain errors for each part of the performance test that employees failed?
   - When time permits, assign a peer for remedial training if any employee performed below standards?

General areas

4. Did you:
   - Speak so employees could hear what you said clearly?
   - Use understandable words?
   - Encourage employee questions?
   - Always answer suitable questions?
   - Always avoid answering irrelevant questions?
   - Remain patient with employees?
   - Reinforce correct employee performance with assurance like "Good," "That's right," and "Fine"?
   - Avoid giving employees unnecessary help?
   - Create an environment that simplified learning by minimizing distractions and provided for evaluation without disrupting training?

Creating a Positive Training Environment

I have outlined the techniques you need to consider to supply your employees with effective and efficient training. It is vital that you prepare, conduct, and evaluate your training effort, but it is equally important to provide a good training environment. You must build an environment that enhances thinking. The creation of a physical space that supports mental energy is critical. The following techniques will help you construct a positive training environment.

Security Technique 166: Create a Controlled Training Space

Creating a learning room begins with finding a space that supplies privacy and allows employees to concentrate without outside distractions. You should ensure that no messages, telephone calls, or other interruptions take place once the training session begins. Business operations must stay "outside" until the training is over.

Take inventory of the existing characteristics of the training space or room. The space should be pleasant so that employees will feel comfortable and will want to train instead of feeling impatient or becoming distracted. Employees in a poor training environment tend to experience a sense of monotony, have low energy, show signs of irritability, and display a strong desire to discontinue the training.

Security Technique 167: Provide Adequate Furnishings, Lighting, and Humidity

Comfortable chairs are critical in your training environment. With performance-oriented training most of the training time will involve the employee practicing the job-related tasks for that session; however, it still will involve sitting, because most employees do their jobs sitting down. Uncomfortable chairs can cause employees to become lifeless, bored, and irritable. Any of these important factors can create a marked reduction in their desire and ability to learn.

The availability of natural light and the level of relative humidity are important considerations in choosing a training environment. Sunlight puts people in a positive mood, enhancing their ability and desire to learn. High relative humidity creates fatigue. The ideal room temperature is 70–73, degrees with the relative humidity ranging from 60 to 70 percent. The temperature and humidity should remain steady.

Quick Reference Guide to Chapter 14: Training Employees Effectively

- The purpose of business-related training is to prepare employees for job performance. Performance-oriented training serves this need best.

- Each part of an employee's job performance should include awareness training in security and asset protection. For example, the floor salesperson needs skill in customer relations, knowledge of the service or product offered, awareness of shoplifting, robberies, and other crimes, and training in how to deter and deal with business crime effectively.

- When you need more than one session to complete the training satisfactorily, set intermediate objectives.

- Your training objective is the key to testing employees' proficiency to perform their jobs in the workplace.

- Use this equation as a reminder throughout your training efforts:

$$\text{Training objective} = \text{Training} + \text{Test} + \text{Evaluation}$$

- Performance-oriented training, whether based solely on security measures or integrated within business and job training, should always be structured so that the training session has three key elements: *task, conditions,* and *training standards.*

- Formulate employee performance-oriented training using three important steps:
  1. Describe the desired results of training.
  2. Prepare to conduct training.
  3. Conduct training to meet your business standards.

- Conduct performance-oriented training in three phases:
  1. Demonstrate your training objective (or intermediate training objective) for that session to employees. This phase may include a brief explanation or lecture that creates the foundation for a demonstration.
  2. Have employees practice the objectives to acquire the proficiency you believe meets your training and job standards.
  3. Test employees through performance of the objective to your standards without assistance, as they would need to do on the job (after practicing it with you coaching).

- Optimize your resources, including people, time, and money.

- Develop training sites conducive to learning.

15

Hiring Security Officers or Consultants

Many businesses need more protection than their internal asset protection systems provide. They hire or contract for the physical presence of security officers. As a part of that process, businesses may also hire the services of a security consultant. If these security measures are not undertaken properly, they can create a minefield of incompetence, incur liabilities for the place of business and business owner, and develop a false sense of security that quickly backfires. I have talked with many business owners and managers who were completely disillusioned with using their hard-won profits to hire security officers and consultants. However, their disappointments usually were caused by a form of blind trust. In this chapter, I will show you how to avoid disillusionment by hiring competent, reliable, and professional security personnel.

Security Officers and Consultants: Some Key Definitions

The first advice I want to give you involves titles. Beware when security agencies or self-employed people bill themselves as "security guards." Their perception of the security that you need and your requirements may differ extensively. There's a big difference between a security guard—usually an obscure, sedentary watchdog type—and a security officer—someone who protects property actively and overtly. Some will claim the difference is only a matter of words, and in certain cases this may be true. However, the term "guard" implies passive protection, while the term "officer" implies that the person thinks, reacts, and has an interest in his or her profession.

How security people or agencies identify themselves will tell you more about how they view their service role than anything they may say to you. What they call themselves may also inform you about their professional capabilities and training. For example, the term "security officer" suggests that the individual or agencies have enough professionalism to think of themselves as "private police," instead of fixtures like store display cases.

My second piece of advice is not to confuse security consultants with security officers. Security consultants can be either generalists or specialists. Their role involves assessing, analyzing, and offering you viable solutions on the basis of their education, training, and experience. If they do not have the proper training or credentials, they will have no idea how to help you solve your asset protection problems.

In recent years, a number of consultants have added the suffix "certified" to their business cards. I caution you against this classification, however, because it does not mean government agency certification. For a sizable fee, private organizations will award certification to those who meet their criteria for excellence in security. Anyone can establish a "certification" program, and I have found that some consultants obtain a "certified" suffix just because it impresses their clients. Even if their certification comes from a reputable agency, as some do, it means only that they met the issuing organization's criteria and may have passed a written test.

Before certification can have any credibility, it must involve stringent government (federal or state) regulation and a standardized program (similar to that required of law enforcement officers). It must also be a prerequisite to a licensing process for security consultants. The dual requirements of licensing and certification will probably not come to pass for years. Therefore, my advice is not to concern yourself with certification but instead to look at the consultant's track record and consider his or her education, training, and experience.

Security Technique 168:
Determine If You Need Security
Personnel

Once you understand that all security officers and consultants cannot be lumped into one category, you should focus on determining if hiring them will be advantageous to your business. First you need to establish the internal business asset protection programs that I have illustrated throughout this book. Only after you have set up these preventives and deterrents, and determined that they do not eliminate the total threat, should you consider hiring security people as "hands-on" supplements.

Usually, security officers can help only in overt threats, such as armed robbery. However, some businesses hire security officers to deter shoplifting, provide access control, patrol parking areas, and perform other functions. If you choose to control shoplifting and forms of other theft in this way, consider keeping your security officer in uniform rather than hiring a traditional plainclothes detective who sneaks around pretending to be a customer. Although house detectives may "catch" a shoplifter now and then, professional or quasi-professional thieves, especially those working in groups, will quickly identify plainclothes officers and frustrate their efforts with decoys. Uniformed security officers, who are easy to spot, have a certain deterring effect by their presence alone (provided they are professional in appearance and demeanor) and can easily cope with confrontations. Also, the uniform deters children and "opportunistic" shoplifters or armed robbers.

You should consider hiring security officers if your business is located in a high-crime area. When your business remains open after dark, and especially when you have large amounts of cash on hand, extra security measures are advisable. You also need to install video equipment and other high-tech devices to help the security officer protect your business assets.

If your business includes manufacturing, warehousing, or similar facilities, you should consider hiring a uniformed security officer to handle outside security, such as entrances and exit gates, storage areas, and other areas vulnerable to theft. If your business has offices or is located in an office building, uniformed security officers can be valuable in controlling access and providing a variety of deterrents.

You should consider hiring a security consultant when you have a specific need. For example, if you plan to expand or remodel your business, a good consultant can help you design built-in security elements that might otherwise be overlooked. A general security consultant can help you to create good loss prevention techniques for your place of business, and assist in training employees and management. The consultant can also act as liaison between you and the sellers of security devices and equipment and oversee their installation. A good consultant can save you many times the amount of his or her fee in the first year.

Security Technique 169: Use Security Personnel Effectively

The effectiveness of hired security officers and consultants depends on how you use them and what you expect from them. I have observed uniformed security officers bagging groceries or dry goods. I have seen uniformed security officers sweeping floors and stocking shelves. If you need added security in your place of business, then require that the officer be totally committed to protecting your business assets.

Consultants will serve your needs best when you personally agree to listen to their advice. Do not expect a "quick fix" with no-cost solutions. Many business owners readily find excuses for not implementing any countermeasures that a consultant suggests—even before the consultant has completed preliminary assessment of their operation.

Some years ago, a client called me with a story of woe about employees and others stealing him blind and begged me to help him solve the problem. Shortly after our first meeting, the business owner "cautioned" me that he did not want to spend much money, did not want to offend anyone, and did not want to create any additional work within existing business operations. In short, he wanted effective security only if it cost nothing and inconvenienced no one. I declined the assignment because this business owner clearly did not want to make any real changes and was looking instead for a "magic solution." I warn you again that there are no easy solutions. Whatever actions you need to take to end opportunities for crime will involve a few changes and the investment of some money.

Security Technique 170: Consider Armed Security Officers

I have always had problems recommending that a business hire "unarmed" security officers. The idea of an unarmed officer seems a contradiction in terms. If you perceive that you need a uniformed security officer in your business, then a clear threat of some type must exist. However, hiring an unarmed security officer may only exacerbate the threat. Would-be thieves will not be deterred, and might even be challenged, by the presence of a uniformed "protector" who does not have the option of using force.

Never hire security officers with the notion that they will or should use *deadly* force to protect property. Armed security officers protect property by deterring crime and directly confronting criminals. They are the buffers between determined thieves and upstanding customers and employees. Their role is to defend themselves and others *against* deadly force—for example, if a confronted shoplifter or thief is carrying a firearm and attempts to use it.

People who put on uniforms that closely resemble those of police officers, and then confront people committing crimes or even attempting access to a building, place themselves in jeopardy at the outset—because they have no way of protecting themselves legally against deadly force. By the same token, if you hire security officers and insist that they remain unarmed, you reduce the deterrent effect of their presence. However, this does not mean that you should just hire a person who carries a gun. You need to make certain that the security officer has significant training in crime prevention techniques, authority and jurisdiction, firearms, and other security issues.

Security Technique 171: Understand the Authority of Security Officers

Security officers have the same legal authority and jurisdiction as you do. When you hire them, they become your representatives. Unless an officer has a special commission or deputization from the local police to arrest suspected criminals, or is an off-duty police officer with departmental authorization to make arrests as a security officer, the only authority he or she has comes under the heading of "citizen's arrest."

The laws governing citizen's arrest are complex and often unclear. They are filled with pitfalls that rely on many fine distinctions, such as the crime being committed, proof of actual presence, and the time and place of the incident. Because of these difficulties, the security officer has to know local laws. Improper actions in making an arrest can expose the officer and the employer to civil suits involving charges of false imprisonment, battery, assault, and malicious prosecution.

Security Technique 172: Know the Scope of a Citizen's Arrest

The power of citizens to make arrests stems from the common law that provides that a private citizen under certain circumstances has authority to act independent of any public authority. Every state has preserved this power of arrest in one form or another, either by court pronouncement or by statutory mandate. Each state has exclusive regulation of the authority to perform a citizen's arrest under its police powers, with citizen's arrest power subject only to the laws of the state in which it happens.

Citizen's arrest powers do not rely on constitutional limits. The due process clauses of the Fourteenth Amendment do not apply to citizens acting on their own. Consequently, the Fourth Amendment (incorporated into the Fourteenth Amendment) imposing limits on seizure does not have application to private conduct. An arrest performed by a private citizen, as authorized by state statute, does not fall within the category of government or state action if the person does not act on behalf of a public authority, such as the police.

Some courts, however, have begun to consider private security officers in a different light. For example, a California district court of appeals ruled that because private store detectives use state law as authority, they are acting as agents of the state in the same way that public police do, and are therefore subject to the same constitutional restrictions.

The view that citizen's arrest is a *right* is questionable. The authority of a citizen (including a security officer) to perform an arrest by authority of the law should more appropriately be termed a *privilege.* There is no statute in the United States that imposes a duty on a citizen (other than a person directed by a peace officer or magistrate) to perform an arrest. Without a legal duty, the limited authority to optionally engage in such conduct creates a privilege and not a right.

The language of various state statutes is broad enough to lead the layperson to conclude that its intended purpose is to encourage intervention. Statutory interpretation in the courts, however, is filled with restrictions and limits that tend to defeat that purpose. The beneficiary of the privilege becomes trapped between the law and its application, and is always confronted by caveat (beware). The result is a conflict between two policies, each one defeating the other: legislative policy encouraging such a form of intervention and judicial policy seeking to limit it.

Contracting for Private Security Officers

Most businesses hire security officers through local security agencies. These security officers come from various walks of life and usually work part time to earn extra income. Although some security agencies have tried to professionalize their services by conducting more stringent screening of their personnel and by providing job-related training and firearms skill development, many just supply a "warm and breathing body" dressed as a security officer.

In recent years, 37 states and many cities and counties across the country have created legislation that calls for licensing of security companies. A few stipulate training, supervision, and preemployment screening as part of the licensing procedure. Only 33 states license full-time security officers and 11 states license part-time officers. Nine states have no regulation of private security officers or agencies.

Security Technique 173: Hire Officers Trained in Prevention and Enforcement

If you decide that your business problems call for security officers on your premises, require certain qualifications from the security agency before making any commitment. Ask about the experience of the officer who will be protecting your business.

Make sure that the uniform and equipment used by the security officer are serviceable, create a good appearance, and comply with state law and regulations. Request that a competent supervisor from the contracted security company check the security officer assigned to your premises at least twice each 8-hour shift. Also check to see that the agency pays the security officers well. Remember, you generally get exactly what you pay for.

Give particular attention to the training that the agency provides. Security officer training should include a minimum 40-hour curriculum with instruction supplied by qualified personnel. The following courses should be part of the training program:

- Orientation
   Role of security officers
   Deportment
   Appearance
- Legal powers and limitations
   Prevention versus apprehension
   Use of force
   Search and seizure
   Arrest powers
- Prevention and protection
   Patrolling
   Checking for hazards
   Personnel control
   Identification systems
   Access control
   Fire control systems

Types of alarms

Law enforcement and private security relationships

- Enforcement activities

Surveillance

Techniques of searching

Crime scene searching

Handling juveniles

Handling mentally disturbed people

Parking and traffic

Enforcing employee work rules and regulations

Observation and description

Preservation of evidence

Criminal and civil law

Interview techniques

- General emergency services

First aid, including CPR

Defensive tactics

Firefighting, temporary techniques

Communications

Crowd control

Crimes in progress

- Special problems

Escorting money, people, and property

Vandalism

Arson

Burglary

Armed robbery

Theft

Drug and alcohol problems

Shoplifting

Security Technique 174: Hire Officers Qualified with Defensive Weapons

When the security officers at your business premises carry firearms or other defensive weapons such as nightsticks and Mace™, make sure that they have received training in the use of each weapon and have a proficiency rating conferred by competent, qualified instructors. The firearms course should include a minimum of 40 hours for a novice and 20 hours for someone experienced with firearms. A separate course, of the same duration, should be taken for each type of firearm the officer might carry on duty (i.e., a handgun or shotgun). The course should be approved for law enforcement department qualification. In addition, security officers carrying firearms must qualify with their assigned or duty firearm at least once each year. Courses for nightsticks and other defensive weapons should be the same as those required by local police departments.

Security Technique 175: Consider Hiring Your Own Security Officers

The advantage of hiring your own security officers is that you control their qualifications, screening, training, and performance, just as you do with your other employees. A word of caution, however: Unless you develop a genuine career-oriented program with excellent salary, benefits, and working conditions, turnover will become a headache for you. For some reason, many business owners feel that they should pay little for security service. However, when salaries are low, services are poor. This is true for security officers as well as other employees.

Even if you pay security people well, you may have problems. In smaller businesses, in-house security officers come to fraternize with other employees and may overlook indiscretions and let standards slide. Contract security officers tend to have more impartiality and objectivity and will be consistent in their enforcement of your business policies and procedures.

Except in large companies, the expense of having in-house security becomes prohibitive. When the cost is combined with the problems of training, uniforms, equipment, insurance, and other aspects, you are likely to find that hiring contract security from outside agencies is preferable.

Choosing and Hiring an Asset Protection Consultant

When you decide that your security problems need the services of an expert, be sure you hire one. Once again, ignore titles and look for experience and results. Rule one of hiring a consultant: "You get what you pay for most of the time." A consultant who spends 30 years getting extensive education, training, and experience doesn't come cheap. If a person presents an extensive background and comes cheap, steer clear; it's a scam. The following techniques will help you screen and then hire consultants to improve your protection of business assets.

Security Technique 176: Determine How Much You Should Pay for a Security Consultant

How much consultants charge depends largely on the geographical area they operate in, the demand for them, and their personal and business expenses. Remember that consultants have to earn their money from time management. They sell you two things: their time and their knowledge. If you want both, expect to pay for it. If you want only advice for little money, buy their books. Most consultants who have genuine expertise write books. Those with less knowledge write magazine articles for trade publications. Still other consultants read the books and articles of their colleagues. Consultants who have little or nothing to offer do not write or read much of anything. These may seem harsh words. However, there are a lot of phony people out there who call themselves consultants but have no qualifications for the job.

With that in mind, expect to pay at least $250 an hour for a good consultant, with a minimum bill of about $5000. For that money, you should expect a personal assessment of your business premises, including interviews and other background activities, followed by a written report on findings and suggested solutions. If you expect more, then also expect to pay more. As noted above, these are "minimum" rates. The more in demand a consultant becomes, the higher the prices will be. Top-ranked consultants may charge from $500 to $2500 an hour, depending on whom they are working for and how complex the security solutions must be. They may also bid a flat fee if the project calls for long-term assessment, analysis, and research.

Security Technique 177: Decide If Consulting Fees Are Worth the Cost

Before you abandon the idea of hiring a consultant because of the cost, consider instead how much you might save. Think about what your losses or potential losses might total and weigh that against the consultant's fee. For example, if your business losses each year total

$100,000, spending half that amount to end the problem would be cost-effective. However, if your losses total $10,000 and a consultant will cost you double that amount, you need some long-term benefit before the consultant's work will pay off. I have performed consulting services for companies that were losing millions each year to shoplifting and other crimes. These losses ended after I assessed and analyzed their problems and offered viable solutions. My fee, although high at first glance, amounted to only a small percentage of the immediate loss and was a trifling sum compared with future losses that were avoided.

Additional benefits you will receive from a good consultant include the uncovering of "hidden" problems (ones you did not know about or blamed on other sources) and the disclosure of potential problems that would create further losses without remedial action. When you weigh these added benefits against the consultant's fee, your cost will probably become negligible.

Security Technique 178: Respect the Security Consultant's Confidentiality

Although some "consultant" types will readily supply you with a list of clients as references, do not expect an expert in asset loss prevention to do so. This rule especially applies if the consultant has a professional standing and solves real problems. The reason is a combination of ethics and confidentiality.

When businesses hire security officers, to other companies and the public the decision is nothing more than common sense and good business judgment. However, when businesses call in security consultants, they usually have serious problems that they do not want to become public knowledge. If financial institutions, creditors, and customers perceive that some major problems have surfaced in a business, people become wary. They speculate on real or imagined difficulties, such as filing for bankruptcy, going out of business, or losing assets pledged as collateral. In the case of a public corporation, such innuendo or suspicion may drive stock prices down, creating even greater losses and financial problems.

Because of the ramifications of their work, top-grade security consultants will rarely if ever reveal the names of their clients and certainly will not discuss client problems. If a security consultant who claims to have the skill of assessing, analyzing, and solving your loss problems supplies you with a list of clients, you should become immediately wary of his or her standing as a professional.

Security Technique 179: Ask the Consultant to Prepare a Proposal

When you call in a top-ranked security consultant, plan to spend enough to pique his or her interest. Ask the consultant for a proposal, but understand that creating a meaningful proposal will take some time. Also, expect to pay for that preliminary assessment.

A typical proposal has three purposes. First, it informs you of how the consultant views your needs. Second, it expresses the consultant's willingness to perform the services for you. Third, it justifies your investment in terms of benefits and advantages. This last point is especially important if you need to consult a board of directors or others before committing your business assets to security services.

Most proposals will contain the following sections, with each explaining its relationship to your specific business.

1. The consultant supplies an introductory overview of your business, including a variety of statistics that create an awareness of typical problems leading to losses. Other information such as the perception of the work the consultant believes will best serve your business and how the consultant has organized the proposal will also be included.

2. The consultant presents an overview of your present situation (after the preliminary assessment) and puts the problem in context. He or she might add specific details found during the assessment and suggest some underlying problems.

3. The consultant lays out objectives and the scope of the work proposed, along with a general statement of what the consultant will and will not do and what you can expect. These points are important because you might ask the consultant to revise the objectives and do more or less than proposed. Generally, your budget will govern the scope of the consultant's work. However, as noted earlier, make your judgment according to the size of your loss problems and how much the consultant's work might save you.

4. The consultant outlines general approaches to solving your business problems. These explanations should predict what the assignment will involve and provide ample detail without committing the consultant or you to an inflexible position.

5. The consultant explains how he or she will work with you and with key people in your business, and discusses how and when you will be briefed on progress and developments.

6. The consultant outlines the expected benefits and results of the

work and lists his or her qualifications, along with those of associates if they are to be part of the project.

7. The consultant gives an estimate of charges per hour, per week, or per month. Depending on the consultant's policy, you may be able to negotiate a flat rate for the work.

8. The consultant presents a timetable for the work, including estimates on planning the assessment, making an analysis, and finding viable solutions to your business asset protection problems.

9. In a review-type summary, the consultant provides an overview of benefits and advantages and other key elements of the proposal.

10. Depending on the scope of the proposed project, the consultant may include his or her professional credentials (and of associates and employees when applicable) and how they apply to your specific business operations.

Remember that consultants run a business, just as you do. Some businesspeople like to become "instant experts" after listening to a professional explain a problem-solving approach in detail. They then try to take on the task themselves, usually with disastrous consequences. Most consultants will speak to you in person, and in their proposals they will emphasize the benefits to be gained. However, they will not specify how they intend to produce these benefits for you, because they do not want to reveal proprietary information. Just as you will not tell a customer all your trade secrets, neither will an experienced security consultant reveal his or her methods.

Security Technique 180: Assess the Consultant's Proposal Accurately

Whenever you receive a formal proposal from a security consultant, assess it carefully. Use this checklist as a guide.

- Does it cover the proposed consulting assignment?
- Does it agree generally with your concerns about the situation?
- Does it express the importance of the matters that the consultant plans to study?
- Does it provide a sufficient explanation of the work intended?
- Does it supply a detailed work plan that clearly describes the consultant's role and the estimated time needed to complete the project?

- Does it explain what role you and your employees will play in the project?

- Does it give you a clear understanding of the access requirements of the consultant, such as the need to review records and interview employees?

- Does it state how the consultant will keep you informed on progress?

- Does it convince you that the consultant has suitable credentials to conduct the work?

- Does it supply you with a detailed discussion of fees and expenses?

Security Technique 181: Obtain a Consulting Agreement Before You Hire

You may negotiate revisions to a proposal before accepting it and deciding to employ a consultant. Once you have a viable proposal, ask the consultant for a written agreement or contract. This is also negotiable. The agreement should have the following sections as a minimum:

I. *Statement of work.*   Describe the performance of consulting services for your place of business and any related matters you may request during the period of the contract.

II. *Payment for services.*   Specifically state what you agree to pay the consultant, and how you must make the payments, such as weekly or monthly, or by invoice or other arrangement.

III. *Period of performance.*   Establish the starting and ending dates for the project, and stipulate that the consultant will produce certain work during that period.

IV. *Not-to-exceed limit.*   Describe the total payment for services within the contract and amounts not to be exceeded without further agreement.

V. *Independent contractor.*   Confirm that the consultant is a business and retains full responsibility for his or her own tax obligations. State clearly that the consultant does not intend to serve as your agent or employee and has no authorization to act for your business.

VI. *Right to act as consultant.*   Ensure that there are no obligations, contracts, or restrictions that would prevent the consultant from entering into or carrying out the provisions of the contract.

VII. *Termination.*   Set out the provisions agreed upon between you and the consultant for terminating the contract.

VIII. *Hold harmless.*   Ensure that the consultant does not hold you and your business responsible for any suits, claims, actions, damages, or any losses whatsoever, resulting from acts or omissions on the part of the consultant and his or her employees, agents, and sub-contractors or resulting from the work and performance within the contract.

IX. *Confidentiality.*   Have the consultant agree to keep confidential all the business information that he or she uncovers and any remedies or findings that have great value to you and your business. Also, ensure that the consultant returns all documents or other materials that you furnished, along with any documents prepared by the consultant regarding your business.

X. *Governing law.*   Confirm that the contract and its provisions come under the laws of the state where the work of the consultant has been contracted.

Quick Reference Guide to Chapter 15: Hiring Security Officers or Consultants

- When you need uniformed security, look for agencies that supply "security officers" instead of "security guards" and ensure that those working in your place of business have the minimum training.

- Consider that an unarmed security officer has no more effect than any other employee in protecting workers and customers from people who might steal and become violent. If you want genuine security and maximum deterrence, choose a uniformed and armed security officer. Make sure the officer is proficient with firearms and well trained in all security duties, including the proper use of force.

- When you hire security officers as employees, remember the added costs of equipment, insurance, and other matters beyond those needed for other employees.

- Security consultants assess, analyze, and supply solutions to business losses created by theft.

- When you hire a consultant, ask for a preliminary assessment (at your expense) followed by a written proposal.

- When you accept a consultant's proposal, ensure that he or she provides a contract or agreement giving the details of what you have mutually agreed upon.

Appendixes

The following appendixes contain valuable techniques and checklists to help you assess physical security, make continuing inspections, and institute a variety of countermeasures. The checklists also supply you with "reminders" of what you should look for in various situations. You will find them important employee training tools to be used as you integrate business asset protection into your overall training program or focus solely on the security functions.

Asset protection needs vary with the type of business, the management team, geographic location, and other factors. I have tried to weave techniques into the checklists that will serve several categories of business operations and at the same time be adaptable to all businesses. In some situations, you might benefit from using more than one of the checklists—for example, if you operate retail outlets resupplied by your warehouse. You might also have several different types of retail or wholesale business operations, operate a financial institution, run a check-cashing facility, or handle significant amounts of money to pay your employees and meet other expenses. In that case, several of the checklists used in combination will prove advantageous to your business asset protection program

Appendix **A**

The Businessperson's Observation, Description, and Identification Guide

Much of your asset protection management must rely on observation, description, and identification of people who target your business to perpetrate their crimes. This appendix summarizes the key techniques of observation and provides a valuable checklist that you can refer to, copy, and use in employee training sessions.

Techniques for Accurately Observing People

Everyone has some distinguishing characteristic or combination of characteristics that sets the person apart from others. These distinctive features are the most important part of a physical description. When your business falls victim to crime, it's important to describe suspicious people so completely and accurately that others (including employees and police) will be able to recognize them.

Deliberate observation of people proceeds methodically. After first noting distinguishing characteristics (such as a limp or being very tall), observe general characteristics, specific characteristics, and changing characteristics. These techniques provide a detailed identification of a suspicious person.

Security Technique 182

Observe general characteristics: sex, race, height, build, weight, and age.

Security Technique 183

Observe specific characteristics: color of hair and eyes, shape of head and face, distinguishing marks and scars, mannerisms, and habits.

Security Technique 184

Observe changeable characteristics: clothing, jewelry, cosmetics, hairstyle, and other features.

Techniques for Accurately Observing Objects

By observing physical objects for later description you may help identify a suspect. Use a pattern of observation, proceeding from the general to the specific.

Security Technique 185

Describe the general type of item, including size and color: "1994 Toyota sedan, four-door, blue."

Security Technique 186

Describe specific distinguishing characteristics: a sunroof in an automobile, a portable radio, a typewriter.

Security Technique 187

Note distinguishing marks, particularly those showing damage or alteration: a broken headlight, a repainted fender, a missing door handle, a scratch on a piece of luggage.

Security Technique 188

Whenever possible, include identifying numbers, markings, or labels: equipment serial numbers, recognizable designs or logos, manufacturer's labels in clothing.

Techniques for Describing General Physical Characteristics

Use commonly accepted and understood terms for describing the general characteristics of a person.

Security Technique 189

Note the sex of a person: male, female, or androgynous (a person not distinguishable as either male or female).

Security Technique 190

Describe the racial or ethnic group of a person as closely as possible: caucasian, African American, Native American, Asian, Arabic, Hispanic, or other. Add your own sense of national and geographic origin to further identify the person—such as "southern drawl" or "foreign accent"—and the region or country if possible.

Security Technique 191

Estimate the person's height. The best technique is simply to compare the person's height with your own. Another technique is to observe an object the person is standing alongside, noting the point where the top of the head is and later measuring the object. For purposes of simplification, estimated height may be stated in 2-in blocks (e.g., 5 ft, 8 in to 5 ft, 10 in).

Security Technique 192

Describe the person's physical build (including posture). In general, is the person large, average, or small (slight)? Specifically, is the person obese (very stout), stout, stocky, medium, or slim (slender)?

Security Technique 193

Distinguish between male and female body types. Female musculature is normally smooth and rounded, while male musculature tends to be angular and distinct. Similarly, for the female the shoulders are characteristically narrower than the hips, while for the male the reverse is true.

Security Technique 194

Estimate a person's weight. As with height, estimate a person's weight by comparing it with your own. Estimates are usually given in 10-lb increments (e.g., 110 to 120 lb; 170 to 180 lb).

Security Technique 195

Determine a person's age. Estimating age can be difficult. Look for lines around eyes, bags under the eyes, and weathering wrinkles on the face and forehead. Some women and men today have cosmetic surgery to make them look younger. People may also disguise their age with makeup. Looking at the hands, neck, and posture will usually help you be more accurate. For convenience, give age estimates in multiples of 5 years.

Security Technique 196

Take notice of the person's complexion: fair, ruddy, sallow (sickly pale), or florid (flushed); light brown, medium brown, dark, or olive; clear, pimpled, blotched, freckled, or pockmarked. For a female, the description should also include makeup habits, such as none, light, or heavy.

Techniques for Describing Specific Physical Characteristics

In the interest of thoroughness and uniformity, you must pattern both observation and description of a person's specific characteristics on systematic lines, normally beginning with the head and progressing downward. The following terminology is standard in describing physical characteristics.

Security Technique 197: Describe the Person's Head

Size and shape of the head are best described as large, medium, or small; long or short; broad or narrow; round, flat behind, flat on top, egg-shaped, high in crown, bulging behind.

Security Technique 198:
Examine the Person's Profile

To best describe a person's profile, mentally divide it into three parts or sections. A viable method is to describe each third in its relationship to the whole and in separate detail. A profile isn't usually as important for identification purposes as the description of a frontal view of the face, unless the profile accents characteristics such as a "hook nose" or scar not readily visible from the front.

Security Technique 199:
Describe the Person's Face

Observe the person's face and determine if it is round, square, oval, broad, or long.

Security Technique 200:
Observe the Hair

A person's hair can be a key part of your description. Designate the color as blond (light or dark), brown (light or dark), red (light or dark), auburn, black, gray, streaked with gray, or white. With bleached, tinted, or dyed hair, describe both the artificial and the natural color if possible. Indicate density as thick, medium, thin, or sparse and hairline as low, medium, receding, or receding over temples.

Describe baldness as complete, whole top of head, occipital, frontal, receding, or the proper combination of types. Hair types include straight, wavy, curly, and kinky. Describe hair texture as fine, medium, or coarse and its appearance as neat, bushy, unkempt, oily, or dry. Hairstyle may be long, medium, or short; parted on left, parted on right, parted in center, or not parted. Widely understood descriptive terms for hairstyles (e.g., crew cut, braids) can be used. Wigs, toupees, and hairpieces should be described carefully and in detail. The attentive observer can often discern a hairpiece from differences in hair texture, color, density, type, or appearance. The arrangement of false hair will often be too perfect, and the edges of the hairpiece will be obvious on close scrutiny.

Security Technique 201: Check
the Person's Forehead

Observing a person's forehead can often provide important details. The general description should be noted as high, medium, or low; sloped forward or backward, straight (vertically) or bulging. Describe forehead width as wide, medium, or narrow (top to bottom); wrinkles or age lines as none, light, deep, horizontal, or curved (up and down). You should also note if the forehead is lighter than the rest of the face, especially during summer months. If it is lighter, and the face is tanned, it could indicate that the person usually wears a cap or hat and works outdoors.

Security Technique 202: Examine the Eyebrows

Eyebrows can reveal much about a person, especially when they are notably different from hair color. Describe the eyebrows as slanting from the center (horizontal, slanted up, slanted down); the lines as straight or arched, separated or connected; texture as heavy, medium, or thin; the hairs as short, medium, or long; style as plucked or penciled. It is important that both the natural and the artificial contour of the eyebrows be noted.

Security Technique 203: Study the Eyes

Describe a person's eyes as deep-set (sunken), medium, or bulging; separation on their face as wide, medium, or narrow; peculiarity as crossed, watery, red, or other; eyelids as normal, drooping, puffy, or red. Include a description of the color of the iris and the person's eyelashes: color; long, medium, or short; straight, curled, or drooping. When describing females, note the color, type, and extent of eye makeup used. Describe eyeshadow as none, light, dark, or irregular.

Security Technique 204: Observe the Person's Eyewear

Note the style of eyeglasses (large, average, or heavy frames); color of frames; type of frames (plastic, metal, or a combination of both); and color of lenses (tinted and the color of the tint, or clear). Additional observations include the type of lenses, such as average, heavy, or bifocals. Contact lenses may prove difficult to observe. Watery eyes and excessive blinking are frequent indications of the presence of lenses. Special types of eyewear, such as monocles, lorgnettes, or pince-nez, should also be noted.

Security Technique 205: Describe the Person's Nose

Observing a person's nose can help confirm an identification later. Note the length of the nose as short, medium, or long; width as thin, medium, or thick; projection as long, medium, or short; base as turned up, horizontal, or turned down. Describe the root of the nose (juncture with the forehead) as flat, small, medium, or large; and the line of the nose as concave, straight, or convex (hooked). Nostrils may be medium, wide, or narrow; large or small; high or low; round, elongated, or flaring. Also note peculiarities, such as broken, twisted to right or left, turned up, pendulous, hairy, or deep-pored.

Security Technique 206:
Observe the Person's Mouth

Observe the person's mouth and describe its size (viewed from the front) as small, medium, or large; expression as stern, sad (corners drooping), pleasant, or smiling. Also note peculiarities, such as prominent changes made when speaking or laughing, twitching, or habitually open.

Security Technique 207:
Describe the Person's Lips

Describe the lips as thin, medium, or thick (as viewed from the front); long, medium, or short (as viewed from the profile); position as normal, lower protruding, upper protruding, or both protruding; color; appearance as smooth, chapped, puffy, loose, compressed, tight (retracted over teeth), moist, or dry. Note a harelip or other peculiarities. For females, describe the color, type, and extent of lipstick applied. Be alert for lipstick applications that change or accent the natural appearance of the lips.

Security Technique 208:
Observe Facial Hair

Note the color of facial hair (mustache or beard), including any difference from the person's hair color, and whether or not it appears newly grown. For males, describe the type of mustache or beard (such as, heavy, long, or short), its style (handlebar, Vandyke, or other unique shape), and the state of grooming (neatly trimmed regardless of length, combed, or unkempt).

Security Technique 209:
Examine the Person's Teeth

Observing a person's teeth can help confirm an identification later. You should look for color (yellow, white, average). Note whether the teeth are receding, normal, or protruding; large, medium, or small; stained, decayed, broken, false, gold, flared, uneven, or missing.

Security Technique 210:
Describe the Chin

A person's chin plays a prominent role in defining appearance. Describe the chin as normal (or average), receding, or jutting (as viewed from the profile); short, medium, or long (as viewed from the front); small, large, pointed, square, dimpled, cleft, or double.

Security Technique 211:
Observe the Person's Ears

Observing a person's ears can contribute to your description. You should describe the size as small, medium or large; shape as oval, round, triangular, rectangular, or other; lobe as descending, square, medium, or gulfed. Note whether the separation from the head is close, normal, or protruding; whether the setting (based on a line extended horizontally back from the outside corner of the eye, which crosses the normally set ear at the upper third) is low, normal, or high.

Also note if the person is wearing a hearing aid. Describe the type of aid (inside the ear, behind the ear, with cord, or cordless), color, and ear in which it is worn. This description can be a key factor in identifying a crime suspect.

Security Technique 212:
Describe the Cheeks

Describe the person's cheeks as full, bony, angular, fleshy, sunken or flat; cheekbones as high (prominent), medium, or receding. Note a female's makeup style for the cheeks.

Security Technique 213:
Observe the Neck

Observing a person's neck can be a natural part of describing the cheeks, since the two sometimes run together, as in a person with heavy jowls. Describe the neck as short or long; straight or curved; thin or thick; Adam's apple as large (prominent), medium, or small.

Security Technique 214:
Examine the Shoulders

Describe the shoulders as small, medium, or heavy; narrow, medium, or broad; square or round, level or one side lower; as viewed in profile, straight, stooped, slumped, or humped.

Security Technique 215:
Describe the Arms

A person's arms reveal identifying points not easily concealed. Describe the arms as long, medium, or short in comparison with the rest of the body. (Average or medium arms end with the heel of the hand about halfway between the hips and the knee.) Note whether the musculature is slight, medium, or heavy. Shoulders and arms combined can reveal whether the person is physically fit, lifts weights, or does heavy work.

Security Technique 216:
Examine the Hands

The hands can reveal age, type of work, and other characteristics of a person's lifestyle. Describe them as small, medium, or large. Note if they appear to be used in labor as opposed to office work; if they show freckles, age spots, veins, weathering, tattoos, or scarring.

Security Technique 217:
Observe the Fingers and Nails

Describe the person's fingers as long, medium, or short; thin, medium, or thick (stubby). Look for deformities, such as missing fingers, disfigured nails, or crooked fingers. Note whether a person has manicured, well-maintained nails. Check whether a woman uses nail polish and the color, and whether the nails are long or short. The general appearance of the hands and fingers usually reveals a person's age.

Security Technique 218:
Describe a Person's Midsection

A person's midsection or trunk is best described in the following terms:

- Overall: long, medium, or short (compared with the rest of the body)
- Chest: deep, medium, or flat, as viewed in profile; broad, medium, or narrow, as viewed from the front; for females, average, large, or full-figured bust
- Back: straight, curved; bowed as viewed in profile; straight or curved as viewed from the rear
- Abdomen: flat, medium, or protruding

Security Technique 219: Describe a Person's Lower Body

Describing a person's lower body is important in the identification process. Use the following descriptive terms.

- Note whether the hips are broad, medium, or narrow as viewed from the front; small, medium, or large as viewed in profile. Keep in mind the differences between the male and female figures.

- Describe the legs as long, medium, or short in comparison with the rest of the body. (Average or medium legs combined with the hips are about half the body length.) Note whether they are straight, bowed (bandy), or knock-kneed; whether the musculature is slight, medium, or heavy.

- Describe the feet as small, medium, or large in comparison with body size. Note all peculiarities, such as being pigeon-toed, flat-footed, or club-footed.

Techniques for Observing Other Physical Characteristics

When you observe others for identification, remember that "nondescript people" have physical attributes that could fit anyone. In such a case, the best clues come from seeking out distinguishing characteristics.

Security Technique 220: Look for Marks and Scars

Note important identifying marks such as birthmarks, moles, warts, tattoos, and scars. Describe them by size, color, shape, and location on the body.

Security Technique 221: Observe a Person's Speech

Describing the tone and manner of a person's speech can add depth to an identification. Note whether the habitual tone is low, medium, or loud; soft or gruff. Describe the manner of speaking as cultured, vulgar, clipped, fluent, broken English with accent (identified whenever possible), or non-English-speaking (language specified when possible).

Identify such peculiarities as stuttering, nasal twang, pronounced drawl, or a mute condition.

Security Technique 222: Examine Dress and Personal Appearance

Since clothing is a changeable characteristic, its value for descriptive purposes is limited. Still, clothing worn during an offense or when a suspect was last seen should be described in detail. Note colors and condition, such as clean, soiled, torn, ragged, greasy, or bloodstained. Overall, what is the style of dress? Is the person neat or untidy, well groomed or unkempt, refined or rough?

Security Technique 223: Check Out Mannerisms and Habits

Often the peculiar mannerisms or traits of a person will be a major part of the description. Stay alert to describe such characteristics as these:

- Peculiarities in moving or walking
- Outward emotional instability, nervousness, or indecision
- Subconscious mannerisms, such as scratching the nose, running the hand through the hair, pulling on the ear, hitching up the pants, jingling keys, or flipping coins
- Facial tics, muscular twitches, and excessive "talking with the hands."

Techniques for Increasing the Accuracy of Observations and Descriptions

Before you can perfect the techniques of developing a description from your observations, you need to develop an awareness of and be able to make allowances for the many factors that influence your perception, interpretation, and retention of details. Such factors may be external or human. Use the following techniques to improve your observations.

Security Technique 224:
Understand External Influences———————————

Witnessing an event from different locations may account for differences in observation. For example, a person who witnesses the event from a great distance may be able to give a good general description of what happened but be unable to explain in detail the characteristics of the people or objects involved. By contrast, a person who witnesses the event at very close range may be able to describe in detail the people, objects, and actions involved but be unable to put together a general picture of what happened. The height of an observer (above or below the subject of observation) also influences interpretation.

You can become a better observer by focusing on specific segments of a situation instead of trying to remember every detail. If the situation becomes confusing, try to select one or two details that seem the most important for later description and identification.

Security Technique 225:
Understand the Influence of
Weather and Light———————————

The effects of weather and light upon human observation create many reliability problems. For example, a witness may claim to be certain of a person's appearance. However, later the identification is discredited because light conditions made it impossible to obtain a detailed observation. Fog, rain, and snow also hinder observation and identification. To overcome these obstacles, concentrate on a single element of a situation. It is better to have one clear and precise detail than a vague overview of an event.

Security Technique 226: Avoid
Distractions———————————

Unrelated but concurrent events may greatly influence a witness's observation of events. For example, an exciting play on a football field will cause a spectator not to observe actions of people sitting nearby. Criminals rely on distractions to avoid being observed and to prevent a clear identification by witnesses. You can improve your ability to observe by getting a sense of your surroundings before distractions take place. For example, at the football game, observe who is sitting around you before the exciting play. You may not witness any criminal act that later takes place. However, you will be able to provide an accurate description of a suspect. In a place of business, don't allow distractions to prevent you from obtaining a full description of suspicious people.

Security Technique 227: Overcome the Time-Lapse Problem

The passage of time between observation and recall can greatly influence a witness's description. The imaginative person may substitute conjecture for incomplete knowledge of an event, particularly if the incident is important to an investigation. With the passage of time, people tend to forget or confuse details of an event. The best way to overcome the "time-lapse problem" is to make notes of your observations. Later, if needed, you can supply an accurate account of what happened and a description of the suspect.

Security Technique 228: Conduct Timely Interviews of Witnesses

It is extremely important to interview witnesses or victims before they have time to adjust their observations, consciously or unconsciously, to other information they may have seen or heard. If an employee tells you he or she witnessed something—a suspicious person in the store or a possible shoplifter who has left the store—get a complete accounting and write down the observations in detail.

Security Technique 229: Understand the Human Factor

Physiological factors, psychological traits, and training all influence human perception. It is within this highly subjective framework that people evaluate and interpret stimuli received by the senses. You must recognize these influences when an employee reports an incident or a suspicious person to you. As noted earlier, it is dangerous to accuse a person of shoplifting solely on the basis of a report from an employee. Some employees have vivid imaginations or tend to exaggerate a situation. Others may unnecessarily cause alarm with their accounts of suspicious people who they believe are about to rob the store. Keep these factors in mind before you decide to react solely on an employee's report.

Security Technique 230: Account for the Influence of Experiences

The evaluation or interpretation of events tends to be colored by people's experiences with similar or related occurrences. Familiar sounds, odors, and tastes will usually be accurately interpreted, while incoming stimuli that have no perceptual reference points may be misinterpreted. Similarly, inaccurate interpretation of experiences in the past will influence later interpretations.

A person of unusually short or tall stature may misinterpret the size of another person. For example, someone 6 ft tall may appear very tall to an observer who is only 4 ft, 10 in height, while the same 6-footer would appear to be of normal height to an observer who is 5 ft, 10 in tall.

Visually impaired people often have sensory perceptions that surpass those of people with normal vision. Thus a blind person may perceive sounds or note details of objects touched that a person with sight will not observe.

Taste and smell are the least reliable of the senses on which to base interpretations. The presence of a strong taste or odor may completely mask the existence of other tastes or odors. For example, the presence of strong cooking smells in a room may lead to a person's failing to note the presence of a more subtle odor of importance to a particular case.

Security Technique 231: Be Alert to Emotional Influences

Temporary or permanent emotional disturbances—such as fear, anger, worry, or mental instability—may weaken a person's senses and lead to inaccurate observation.

For example, a robbery victim may experience such fear of a weapon used by a perpetrator that his or her recollections of the incident are limited to an exaggerated perception of the size of the weapon. The victim probably cannot accurately describe the perpetrator or the type of firearm. Similarly, strong emotional feelings (positive or negative) about someone may cause a witness to see only the actions of that person, to the exclusion of the actions of others involved.

Security Technique 232: Guard Against Prejudice

Occasionally, although a witness has an objective view of an event, prejudice against a social group or race brings forth an inaccurate interpretation. For example, a person who has formed a dislike for police or authority figures may unwittingly allow that prejudice to affect his or her interpretation of the actions of a security officer.

Security Technique 233: Understand Personal Limits

Defense mechanisms, survival instincts, and other factors place limits on what an observer can and will divulge. For example, a victim or witness may give only those details that he or she thinks are important. Or the witness may intentionally withhold information so as not to become involved or to avoid personal inconvenience. Sometimes language inadequacies or lack of expressive ability prevent an observer from giving an accurate description.

Criminal Identification Checklists

Criminal identification begins with a quick, overall description of a suspect (see Fig. A-1). Such a description can be broadcast immediately to local police units. Follow up with a detailed account of specific features, mannerisms, and habits, as well as overall demeanor. The following checklists can serve as your guide.

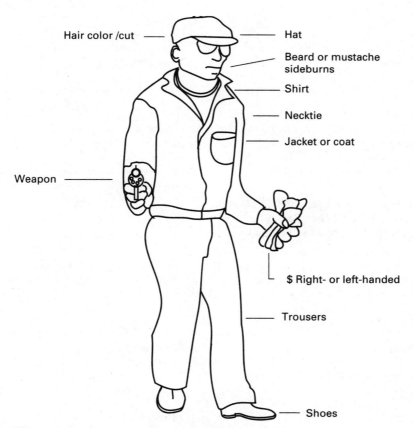

Figure A-1. Criminal description sheet.

Security Technique 234: Use a
Fast General Description of a
Suspect

Male_____	Female _____
Caucasian _____	African American_____
Asian _____	Other _____
Height _____	Weight _____
Nationality _____	Complexion_____
Build _____	Posture_____
Eye color_____	Eyeglasses _____
Eyes alert _____	Eyes normal _____
Eyes droopy _____	Tattoos _____
Marks, scars _____	Age estimate _____
Hat_____	Hair color _____
Mustache _____	Beard _____
Sideburns _____	Haircut style _____
Shirt_____	Necktie_____
Jacket_____	Coat _____
Suit _____	Blazer _____
Trousers _____	Shoes _____
Weapon_____	Other _____

Security Technique 235: Use a
Detailed Checklist for a
Suspect's Characteristics

Sex

☐ Male   ☐ Female

Race

☐ African American   ☐ Caucasian
☐ Asian   ☐ Native American

Head Shape and Size

☐ Round   ☐ Flat on top
☐ Long   ☐ Broad

- ☐ Narrow
- ☐ Square
- ☐ Small

- ☐ Flat behind
- ☐ Bulging behind
- ☐ Egg-shaped

Hair and Haircut

- ☐ Long
- ☐ Full
- ☐ Bald
- ☐ Kinky
- ☐ Straight
- ☐ Red
- ☐ Black
- ☐ Brown
- ☐ White
- ☐ Mussed

- ☐ Short
- ☐ Thinning
- ☐ Curly
- ☐ Crew cut
- ☐ Wig/toupee
- ☐ Hazel
- ☐ Blond
- ☐ Gray
- ☐ Neatly combed
- ☐ Other _____

Complexion

- ☐ Normal
- ☐ Blotched
- ☐ Pale
- ☐ Acne
- ☐ Medium

- ☐ Scarred
- ☐ Pockmarked
- ☐ Florid
- ☐ Fair
- ☐ Dark

Eye Specifics

- ☐ Blue
- ☐ Black
- ☐ Green
- ☐ Large
- ☐ Medium
- ☐ Cocked
- ☐ Sunken
- ☐ Albino
- ☐ Glassy

- ☐ Brown
- ☐ Hazel
- ☐ Crossed
- ☐ Squint
- ☐ Small
- ☐ Bloodshot
- ☐ Multicolored
- ☐ Protruding
- ☐ Blinking

- ☐ Piercing
- ☐ Close-set

- ☐ Dull
- ☐ Bags under

Eyeglasses

- ☐ Plastic
- ☐ Plain lens
- ☐ Thick
- ☐ Bifocals

- ☐ Wire frames
- ☐ Dark lens
- ☐ Normal
- ☐ Monocle

Eyebrows

- ☐ Heavy
- ☐ Dark
- ☐ Bushy
- ☐ Average

- ☐ Thick
- ☐ Light
- ☐ Thin
- ☐ Other _____

Nose

- ☐ Large
- ☐ Wide
- ☐ Narrow
- ☐ Turned-up
- ☐ Short
- ☐ Pug

- ☐ Small
- ☐ Medium
- ☐ Hooked
- ☐ Long
- ☐ Flat
- ☐ Pointed

Mouth Size and Features

- ☐ Small
- ☐ Large
- ☐ Normal
- ☐ Dry lips
- ☐ Lipstick

- ☐ Medium
- ☐ Thin lips
- ☐ Thick lips
- ☐ Moist lips
- ☐ Other _____

Beard and Mustache

- ☐ Mustache/color_____
- ☐ Beard/color _____
- ☐ Long
- ☐ Trimmed
- ☐ Full

- ☐ Short
- ☐ Unkempt
- ☐ Other _____

Teeth

- ☐ Normal
- ☐ Large
- ☐ Cracked
- ☐ Irregular
- ☐ Braces
- ☐ Stained
- ☐ Gaps

- ☐ Missing _____
- ☐ Small
- ☐ Protruding
- ☐ Wide-spaced
- ☐ Close-spaced
- ☐ Fillings
- ☐ Other _____

Ears

- ☐ Small
- ☐ Large

- ☐ Medium
- ☐ Other _____

Chin

- ☐ Jutting
- ☐ Pointed
- ☐ Square
- ☐ Dimpled

- ☐ Receding
- ☐ Normal
- ☐ Double
- ☐ Short

Neck

- ☐ Short
- ☐ Thick
- ☐ Other _____

- ☐ Long
- ☐ Slender

Shoulders

- ☐ Small
- ☐ Broad

- ☐ Medium
- ☐ Stooped

Hands and Fingers

- ☐ Small
- ☐ Large
- ☐ Stubby
- ☐ Rings _____
- ☐ Missing _____

- ☐ Average
- ☐ Short fingers
- ☐ Long fingers

Trunk of Body

Chest

☐ Large   ☐ Average

☐ Narrow   ☐ Sunken

Back

☐ Straight   ☐ Curved

☐ Humped   ☐ Bowed

Waist

☐ Small   ☐ Average

☐ Large   ☐ Protruding stomach

Hips

☐ Broad   ☐ Average

☐ Thin   ☐ Other _____

Legs

☐ Long   ☐ Short

☐ Average   ☐ Bowed

☐ Straight   ☐ Muscular

Feet

☐ Small   ☐ Average

☐ Large   ☐ Other _____

Marks/Scars/Tattoos

☐ Tattoos _____

☐ Scars _____

☐ Marks _____

Speech

☐ Normal   ☐ Rough

☐ Coarse   ☐ Lisp

☐ Whisper   ☐ Strained

☐ Soft   ☐ Harsh
☐ Deep   ☐ Average
☐ High   ☐ Raspy
☐ Vulgar   ☐ Accent
☐ Cultured   ☐ Drawl
☐ Articulate
☐ Explain _____

Dress and Appearance

☐ Hat or cap, kind and color

_____

    ☐ New   ☐ Old   ☐ Faded
☐ Blouse or shirt, kind and color

_____

    ☐ New   ☐ Old   ☐ Faded
☐ Dress or skirt (female), kind and color

_____

    ☐ New   ☐ Old   ☐ Faded
☐ Purse/other (female), kind and color

_____

    ☐ New   ☐ Old   ☐ Faded
☐ Trousers, kind and color

_____

    ☐ New   ☐ Old   ☐ Pressed
☐ Shoes, kind and color

_____

    ☐ New   ☐ Old   ☐ Shined
☐ Shirt   ☐ Sweater   ☐ T-shirt
Color _____
    ☐ New   ☐ Old   ☐ Worn
☐ Necktie   ☐ Bandanna   ☐ Ascot
Color _____
    ☐ New   ☐ Old   ☐ Worn

☐ Suit   ☐ Blazer   ☐ Jacket

Color _____

☐ New   ☐ Old   ☐ Worn

Prosthetic Devices

☐ Crutches   ☐ Hearing aid
☐ Cane   ☐ Braces
☐ Other _____

Overall Appearance

☐ Neat   ☐ Mannered
☐ Dirty   ☐ Rumpled
☐ Groomed   ☐ Cologne
☐ Unkempt   ☐ Crude
☐ Clean

Overall Build

☐ Large   ☐ Average
☐ Small   ☐ Tall
☐ Short   ☐ Straight
☐ Stooped   ☐ Agile
☐ Clumsy   ☐ Muscular

Mannerisms and Habits

☐ Calm   ☐ Nervous
☐ Twitches   ☐ Facial tics
☐ Smokes   ☐ Chews tobacco

Subconscious Mannerisms

☐ Pulling an ear
☐ Scratching or pulling the nose
☐ Jingling keys or change
☐ Hitching up pants
☐ Running hand through hair
☐ Other _____

Physical Security Checklist for Retail Stores

Security Technique 236: _____

1. At the close of each business day, do you or your managers inspect the premises to ensure closing and securing of all windows, doors, safes, and files?

2. Do you or your managers ensure that no one remains hidden in rest rooms, stockrooms, or other areas of the store?

3. Have you installed an intrusion detection device in the store, and do managers closing the business set the alarm as the last task before leaving?

4. Do you keep service entrances locked from the inside when they are not being used for authorized movement of stock and merchandise?

5. Do you keep a strong light burning over your store safe or other important storage cabinets during nonbusiness hours?

6. Is your safe visible to security or police patrols passing by the store?

7. Does any part of the store lend itself to pilferage or shoplifting opportunities?

8. Have you supplied adequate training to your employees regarding shoplifting prevention techniques and how to handle detected shoplifters?

9. Do you use a tally-in, tally-out system for checking supplies, merchandise, or products against shipping documents?

10. Do you conduct a specially designed inventory of stock at the close of each business day?

11. Do you closely audit shortages of high-value items each day?

12. Do you have a plan of action when shortages appear?

13. Do your sales slips show the serial numbers of items when appropriate?

14. If not, will recording serial numbers enhance your inventory control?

15. Do you keep valuable merchandise such as watches, cameras, and jewelry in secured storage areas apart from normal stock?

16. Do you ensure that high-value items have adequate protection from pilferage and shoplifting?

17. When your service doors stay open, do you post an employee there to prevent unauthorized entry or exit and to safeguard assets from theft?

18. Do you control which employees have access to the stockroom?

19. When unauthorized people enter the stockroom or other storage areas, do you have an employee accompany them?

20. Do you or your managers routinely inspect garbage or trash before it leaves the store to make sure it does not contain merchandise or material that is being removed without your approval?

21. Do you require your employees to store all personal packages, parcels, briefcases, and similar items in a secure area close to the employee entrance and exit?

22. Do you have a rigid policy that all employees leaving the store must first check with the manager?

23. Do you allow employees to carry their bags or purses into the sales area instead of securing them in a personal locker within the store?

24. Do you have employees enter and depart the store through one door supervised by you or your managers?

25. Do you ensure that employees buy items from the store after their shift, and do you have a manager or supervisor ring up the sale?

26. Do you have tight controls on cash registers to ensure accurate accountability and detect shortages?

27. Do you make sure that managers verify and record cashier over-ring errors?

28. Do you periodically remove excess money from cash drawers and supply the cashier with a receipt?

29. Do your cash register displays allow the customer to see the amount rung up?

30. Do you periodically take surprise counts of the cash drawers?

31. Are all cash register drawers equipped with locks and keys?

32. Do you ensure that all cash drawers stay locked when unattended?

33. Do you bag and count change at the close of business each day?

34. Do you call for ample identification and other protective measures when people pay with checks, money orders, traveler's checks, and credit cards?

35. Do you refuse to accept postdated checks?

36. Do you have adequate protective systems for cashiers, management offices, and other areas that handle money?

37. Do you have any business safe weighing less than 1000 lb secured to the floor?

38. Do you periodically change the combination to the store safe and install a new combination whenever a turnover occurs in management staff?

39. Do you have adequate control over fitting rooms (in stores selling clothing)?

40. Do you limit controls delegated to management staff?

41. Do you conduct routine, surprise, and spot inventories of the stores and departments, including the stockroom?

42. Do you have a plan for investigating and fixing responsibility for inventory shortages?

43. Do you provide regular, dynamic employee training that includes some material on asset protection awareness in each session?

44. Do you have a secure method for transferring large sums of money from the store to banks?

45. Do you make intelligence information available to all cashiers and other employees about known or suspected fraudulent check and credit card passers, shoplifters, and other criminals targeting businesses?

46. Do you control accountability for waste and damaged merchandise, products, or expendable items?

47. Do you have strict controls on expendable items in your store, including items directly associated with the daily sales as well as administrative and other supplies?

48. Do you control access to and usage of petty cash funds?

49. Do you maintain strict controls on store or business vehicles?

50. Do you maintain tight controls on business information, including access to it?

Appendix **C**

Physical Security Checklist for Retail Grocery Stores

Retail grocery stores and related facilities have physical protection requirements that set them apart from other retail stores. The following techniques are designed to develop the "security habit" among grocery employees, managers, and supervisors. Three key areas need attention:

- Protecting foodstuffs
- Creating a safe environment
- Eliminating security weaknesses

Security Technique 237: Establish Controlled Areas

Establishing controlled areas for the security of food items will improve the total protective posture of the store and related facilities and increase security consciousness among employees.

Although the use of controlled areas may cause some slowdown in operations and may inconvenience employees, it will help you identify security weaknesses.

Controlled food store areas include:

- Parking areas for incoming shipments of food items
- Loading and unloading areas, including the direction of travel to specific points within a warehouse or to a sales outlet
- Surveillance by physical or electronic methods
- Control of employees, patrons, vendors, and others

Security Technique 238: Safeguard Store Construction Areas

Doors to the store must be adequately constructed to ensure asset protection. Exhaust outlets, air-conditioning and heating ducts, and other openings of more than 96 square inches should have bars or grillwork or be covered with a chain-link material. Crawl spaces beneath the store should be inspected and secured to prevent access from under the building. Construct, remodel, or fortify the building to deny access through walls, roofs, and other areas.

Security Technique 239: Protect Service Entrances

Design the service entrance to the store so that supervisors can observe entry and departure. The entrance should be secured to allow opening from the inside only. Service doors must stay locked until the precise opening hours and be unlocked only by designated managers or supervisors.

The store entrance should remain far enough from the cash registers to allow observation of pilferage. Set up guard rails to ensure that customers entering to purchase items pass near the cashier stations.

Service entrance doors to the warehouse or stockroom must remain locked, except during loading and unloading operations. The doors should be kept under close observation when not secured. Outside door hinges should be of the lock-pin variety or be welded to prevent their removal. Install hasps on padlocks for the same purpose.

Security Technique 240: Observe Customers

The store layout should lend itself to maximum observation of attempted employee pilferage or patron shoplifting. Place circular mirrors at strategic points within the shopping areas to enable employees to observe dead spaces and partially hidden areas. Design shopping aisles for maximum flow and minimum shoplifting.

You should also protect storage bin areas from possible tampering. Provide adequate security for cardboard boxes, bottles, cans, and other items that have monetary value when contracted by weight (recycling).

Security Technique 241:
Monitor Incoming Products

Use a tally-in, tally-out system to check items and supplies received or shipped against the appropriate shipping documents.

File all incoming shipping documents and periodically review and audit them. Make sure that incoming items are continually observed during unloading, delivery, verification, and storage. Vendor representatives and other nonemployees should not have access to stockrooms and storage areas unless they are supervised by a responsible employee.

Security Technique 242:
Control Meat Disposal

You or a competent manager should witness the destruction of meat supplies considered unfit for human consumption to prevent good meat products from being diverted from the store under the guise of unfit meat.

Security Technique 243:
Conduct Inventories

Conduct periodic unannounced inventories of randomly selected but popular products in sections of the store. Compare the results with other records. Make sure that all employees know you use this control procedure.

Security Technique 244:
Establish Cash Register
Procedures

Excess cash is tempting. The amount of cash in the change fund for each register should not exceed the amount necessary to carry out business activities. Remove excess cash from the registers regularly and give the responsible cashier a receipt.

Each cashier in turn should sign a receipt for either the cash register or the change fund. You or a manager should verify and reconcile register tapes and cash receipts at the end of each business day.

Display the cash register so that customers and others can see the amounts rung up, including the total. Establish a sound check acceptance and check-cashing policy. Finally, make sure that manufacturer's and local discount coupons are used as intended. Install safeguards to prevent patrons or employees from using coupons as a means of obtaining cash.

Security Technique 245: Create Lock and Key Control Systems

You must keep tight control over locks and keys to the store and entrances. Issue keys only to those employees with a proven job-related need to have them. Keys are not for convenience.

Keys to all locks within the store should be issued on a need basis from a key locker and returned immediately or no later than at the close of business. Do not allow employees to pass on the keys to others. Have other employees who want the same keys request them from the manager of the key locker. Maintain a sign-out, sign-in log for issuing and receiving the keys.

Do not allow employees to remove keys from the store. Only managerial employees who have a job-related need for the entrance door should remove that one key from the premises. If any management turnover occurs, change the lock immediately.

Tightly control and restrict knowledge of combinations to locks. Issue combinations to safes only to those employees who need access. When practical, rotate locks between stores or within the store.

Physical Security Checklist for Warehouse Activities

Security Technique 246

1. Is there a procedure in effect to ensure strict accountability of all property?

2. Do you conduct periodic audits?

3. Do you conduct comprehensive physical inventories regularly?

4. Do you account for fixed and real property?

5. Do you maintain stock records and bin cards?

6. Do you have systems of verification through records of incoming stock?

7. Do you have adequate control over damaged items?

8. Do you have a specific and secure procedure for the receipt of incoming property?

9. Do you require delivery personnel to produce a bill of lading or other listing of the delivery?

10. Do you ensure an accurate tally of incoming property before accepting it?

11. Do you check delivery records against requisitions or purchase orders?

12. Do you designate responsible people to receive property?

13. Do you check incoming shipments carefully for signs of pilferage and damage?

14. Do you keep shipping and receiving platforms free from trash, and do you stack shipments neatly for proper observation and accurate counting?

15. Do you keep unauthorized people from storerooms and storage areas?

16. Do you maintain a personnel access list?

17. Do you protect warehouse contents against pilferage?

18. Do you maintain adequate protective measures for open storage property?

19. Do you keep open storage property properly stacked, placed within, away from, and parallel to perimeter fences, in order to give employees, security, or police patrols an unobstructed view?

20. Do you supply locker and "break area" facilities for employees away from the warehouse storage areas?

21. Do you maintain a secure place for cases of damaged property to prevent pilferage?

22. Do you restrict or control employees from carrying packages or other containers into the warehouse work areas?

23. Do your parking regulations ensure that vehicles will not be used to remove items from the warehouse?

24. Do you prevent trash collectors from entering the warehouse, and do you have procedures for removing all trash to a central outside collection point?

25. Do you supervise employees as they remove trash from the warehouse?

26. Do you have an effective procedure for the issuance and shipping of property?

27. Do you require receivers of property to sign for the goods?

28. Do you have signature cards on file for people authorized to receive property from the warehouse?

29. Do you have a satisfactory system of responsibility for the various warehouse activities?

Physical Security Checklist for Finance and Accounting Centers

This checklist or parts of it are applicable in office environments, financial institutions, and similar business operations.

1. Do you have adequate security measures during business hours to discourage and prevent armed robberies? (Eliminate the opportunity.)

2. Have you taken adequate measures to prevent unauthorized access during nonbusiness hours?

3. Have you coordinated your security measures and procedures with local police agencies?

4. Do you have adequate facilities for storing and safeguarding funds and documents?

5. Do the safeguards used during normal operations prevent loss, substitution, or pilferage of monies and documents? (Remember, information might also have considerable value.)

6. Have you ensured that vaults and safes remain inaccessible to unauthorized people?

7. Do safeguards such as railing or counters prevent unauthorized people from entering office working areas?

8. Have you arranged your money-exchanging windows to prevent unauthorized access to cash located there?

9. Do you have a prudent deposit procedure to ensure that no excess funds remain in the facility?

10. Do your internal office procedures supply controls for undelivered and returned checks?

11. Do you have a central point for receiving, holding, and making final disposition of checks, with responsibility charged to specific people?

12. Do you provide cashiers with a separate working space, or properly enclosed cage or room with a window, for paying and receiving?

13. Do cash drawers have key locks or a small drop safe (when applicable) to safeguard monies, vouchers, and documents during the temporary absence of cashiers or clerks?

14. Does each person in your facility who handles cash have a separate and secure receptacle for the monies?

15. Do your cashiers or clerks sign receipts for the money entrusted to them, and do you supply receipts for the money returned?

16. Do you hold regular, unannounced verifications of cash on hand?

17. Do you have adequate controls and procedures for making cash payments?

18. Do you supply your cashiers with current information about people committing fraud at an institution, facility, or office like yours?

19. Is the counting of money by a cashier or employee verified by a manager at the close of business or at the end of a shift?

20. Are cashiers, clerks or employees who handle funds adequately bonded?

21. Do you keep cash or other negotiable instruments entrusted to employees within the limits of their bonds?

22. Do you keep blank business checks securely locked in a safe, and do you have adequate accountability procedures for them?

23. When you receive an order of blank business checks, do you examine the shipping cartons and audit the check serial numbers to ensure that none are missing?

24. Do you have adequate procedures for safeguarding spoiled and voided checks, money orders, stocks and bonds, savings bonds, or similar instruments?

25. Do you conduct periodic audits and inspections of blank, spoiled, and voided business checks?

26. Do you have adequate procedures for preventing loss when business checks are mailed or delivered?

27. Do you have adequate control and accountability for keys and combinations throughout your office, institution, or business facility?

28. Do you replace locks when a key is lost or missing, or when there's a change in personnel that affects access authority?

29. Do you keep a reserve key or combination to each critical lock in a sealed envelope in a safe or vault?

30. Do you mark the envelopes to detect any unauthorized opening?

31. Do you have procedures to open or take control of a cashier's drawer and safe when the responsible employee is not present (because of sudden illness, suspected crime, or other reasons)?

32. When you must take control of a cashier's drawer, do you have two or more disinterested parties witness the opening and counting of money?

33. Do you have witnesses prepare and sign an affidavit about the contents and monies after such actions as 31 and 32 above?

34. Do you place some shielding around the dials of vaults or safes so that others cannot observe the operation of the combination?

35. Do you change combinations of vaults and safes at least every 6 months?

36. If you use security officers, do they understand their role and instructions properly?

37. Do security personnel have adequate training for their position?

38. Does your security department have adequate communications to summon police in an emergency situation?

39. Do you have adequate safeguards and prevention plans when your funds are in transit?

Appendix F

Checklist for Preventing White-Collar and Economic Crime

The following checklist presents controls and preventive measures against specific frauds that target businesses. The suggestions provide a foundation on which you can build your security program and later adapt it when you receive added information from business organizations or the police. You can tailor these techniques to fit the precise needs of your business and local conditions.

Preventive Measures to Combat Deceptive Practices

Dishonest competitors and unethical employees, operating independently or in collusion, pose a serious threat to your company's financial security. Deceptive practices against your customers can also undermine your business operations—creating both immediate financial setbacks and loss of future business. Use the following security techniques to ward off deceptive practices, from both inside and outside your organization.

Security Technique 248:
Eliminate Losses from
Bankruptcy Fraud

Remain alert to scam operations by those you do business with. Early-warning signals include changes in management, the overly easy sale, and orders involving goods unrelated to your customer's business.

Review the dollar cutoff limit when you order a credit check and when you routinely ship the goods. Bankruptcy planners often depend on high cutoff points. Also review your business policy on rush orders. When you receive a rush order from a new account, do not omit a credit check and other verification searches.

Seek the assistance of a credit-rating organization. Make certain, however, that the rating agency does not provide you with unverified information supplied by the suspect company itself. When you hear of a change in management at a customer business, find out whom you now must deal with.

Do not place too much reliance on sales commissions. Doing so tempts the salesperson to book orders despite strong suspicions about a customer.

Instruct your shipping personnel to report destinations that seem incompatible with the product ordered. For example, a shipment of television sets destined for, or rerouted to, a machine shop should send up a warning flag. When you ship to an unverified risk customer COD, instruct delivery drivers not to accept checks. Even certified checks can have a counterfeit "certified" stamp.

Security Technique 249:
Eliminate the Opportunity for
Bribery, Kickbacks, and Payoffs

Maintain a strict policy about gifts that employees may accept from other businesses or individual customers. Acceptance of valuable gifts might jeopardize your company's integrity and image and tempt your employees to favor the person or business offering the gift.

Separate your business receiving operations from the purchasing department so that buyers will not be tempted, through collusion with receiving personnel, to accept short deliveries for kickbacks.

Whenever practicable, use a competitive bid method of purchasing. Make sure that a knowledgeable manager outside the purchasing department reviews bids and inspects the quality of incoming goods to determine if the seller meets your specifications.

Ensure that your vendors know they will receive a hearing if they have any complaints about the fairness or impartiality of your business purchasing procedure. Have buyers and other purchasing employees submit monthly reports on gifts, gratuities, and other forms of consideration received from vendors.

If possible, rotate your business to different suppliers so that the purchasing department does not become overly close to one supplier. When applicable, create business policies that maintain an arm's-length relationship with unions and sources of financing.

Set high standards of performance. Buyers whose lines sell in your store should receive rewards, while those whose purchases too often wind up in a bargain basement need to be investigated.

Often, a bribe payment or payments will show on books and records as "commissions," and honest employees sometimes become duped into believing the payments are legitimate. An internal alarm system should sound:

- When payments of "commissions" do not have the usual paperwork, such as canceled checks and invoices (often bypassed with illegal payments)
- When the alleged commissions do not fall within parameters of recognized trade practices
- When payments come through banks not used in your normal business transactions

Security Technique 250: Prevent Fraud and Deceptive Practices Against Consumers

Collective action is the most powerful weapon to combat fraud and other illegalities directed at consumers by unethical employees. The following checklist of questions will help you "harden" your business operations:

- Are your sales goals and other performance standards so difficult to achieve that they encourage your sales staff to use deceptive practices against consumers or other business customers?
- Do you evaluate complaints from customers and view them as possible indicators of deceptive practices by your employees?
- Do you instruct your employees to report signs of consumer fraud, such as price fixing in other competitive businesses?
- Do your salespeople depend so much on commissions that they could be tempted to use deception to get an order?

- Do you encourage feedback from consumers or business customers?

- Do you routinely take a firsthand look at your business operations and compare printed policy and procedures with day-to-day activities?

- If you hire a debt collection agency, does that business keep its operations legitimate and aboveboard?

Preventive Measures to Thwart Computer-Related Economic Crimes

Computer-related crimes account for billions of dollars in business losses annually. Designed to speed up and simplify business operations, computers can also (in the hands of unethical users) rapidly precipitate a business's demise. Control is essential to protect your company's assets. Here are the steps to take.

Security Technique 251: Isolate the Computer Function

Separation of responsibilities—for example, by not allowing programmers to serve as computer operators and vice versa—is a good place to start basic control. Dual control of functions will also help you avoid proliferation of narrow, uninteresting jobs. For example, rotate programmers and operators among different programs and machines.

Keep your computer operations independent of other business departments. Computer department personnel should not have access to such business assets as cash and inventories and should not be in a position to authorize checks, purchase orders, shipping documents, and the like.

Restrict the use of computer equipment, particularly after hours and on weekends. Afterhours processing should have the same strict controls and procedures that apply during the normal workday. Encourage computer workers to take their vacations during a period covering month-end activities.

Transactions listed as exceptions because they did not pass any control points should be investigated and resolved promptly. Ensure that you log unexplained errors, stoppages, or interruptions, and take corrective action. Regularly review physical and data security measures and conduct frequent audits. Most important, allow access to computer facilities and equipment only on a need-to-know basis.

Security Technique 252:
Institute Program-Related
Controls

Ensure that programmers supply written, not oral, instructions to computer operators. Programs must contain a statement of ownership. Identification will help you deter program thieves or facilitate their later prosecution.

Controls written into programs should not be indiscriminately overridden. Exception reports should reflect overrides. Whenever possible, separate the responsibilities of writing, authorizing, modifying, and running programs. Also separate program debugging from production activities.

Build threat-monitoring into programs to detect suspicious deviations from standards or normal patterns. For example, if your payroll should be $150,000 but the total comes to $180,000, the program should create a special report.

Responsibility for maintaining programs that are susceptible to fraudulent manipulation should be divided among two or more employees. Keep an audit trail of changes to programs, and consult computer auditors about appropriate tests and checks to integrate into your application programs.

Security Technique 253:
Control Data Input, Output,
Tapes, and Disks

Institute controls over all documents sent to the computer center for input processing. Compare the number sent with the number received. Ensure that only the originating department has authority to make corrections on source documents for input processing. Make sure that vendor checks, payroll checks, and other negotiable items, as well as important forms such as purchase orders, have sequential numbering.

A good way to control both the number of records processed and the accuracy of the processing is to develop a totaling system. Determine totals of purchase order numbers, stock item numbers, account codes, and so on. Compare those totals throughout the processing procedures. Omissions, duplicate processing, and other errors will be detected. Sequence checking can reveal numbering gaps (as might happen in a check-writing run) and locate duplicate numbers.

Retain source records for a sufficient time so that they can be related to output documents when necessary. Place output in secure storage until it is routed to authorized recipients. Keep your tape and disk library staffed or locked, and establish clear accountability for library maintenance.

Security Technique 254:
Safeguard Data
Communications

Use machine-readable cards or badges that identify terminal users and give them controlled access to your systems. Control the cards by collecting them at the end of each work shift and reissuing them at the beginning of the next. Periodically change entry passwords and security codes that identify the terminal and the user respectively.

In high-security situations, install fingerprint readers and hand-dimension readers to identify terminal users seeking access to your system. Safeguard the privacy of information transmission by installing scramblers and cryptographic devices. Have users sign off when they leave the terminal, noting when they will return to an active status. This procedure will isolate anyone involved in a scheme to access the system illegally during inactive periods. As an added precaution, in case users (knowingly or inadvertently) fail to sign off, program your computer to disconnect terminals after a specified period of inactivity. Illegal users cannot access a disconnected system.

A variety of devices and procedures are available to limit terminal users' access to specified files. For example, allow users to read certain files but not to modify them. Also consider lock words that users can change at any time to protect the file from viewing by others.

Ensure that your system records unsuccessful attempts by terminal users to gain entry to the computers. For example, users may enter invalid codes or request access to files without authorization. In such cases, the system should disconnect automatically.

Preventive Measures to
Combat Illegal Competition

To combat competitive illegalities aimed at those in business and the professions, adopt the following security countermeasures.

Security Technique 255:
Understand the Advance-Fee
Fraud

Do not seek business loans through brokers that deal with offshore or other foreign lenders unless a detailed investigation shows the legitimacy of those sources. If the institution offers you a Dun & Bradstreet report claiming to substantiate its financial statements, obtain another

D&B report independently and compare the two. Do not deal with any lender that will not provide an audited financial statement, certified by an accountant whose integrity those in the profession will confirm.

Verify that the executives listed as board members of the lending institution really occupy such positions. If the lender represents an insurance company unfamiliar to you, check out the company with your state's insurance commissioner. If the broker is willing to put an advance fee in an escrow account until you receive the loan, verify that the bank where the fee will be deposited is not a dummy corporation created by the lender.

Security Technique 256: Recognize the Ponzi Game

In the 1970s Charles Ponzi gained notoriety for a unique investment swindle in which early investors were paid off with money contributed by later investors in order to encourage bigger and bigger risk.

Do not place your business investment funds with someone whose reputed financial wizardry is based on hearsay. You need documented proof—especially proof from investors who are not current clients. Any scheme that promises a fantastic return at minimal risk should become suspect immediately. Verify claims made by the investment specialist that handles the funds of well-known personalities.

Security Technique 257: Beware of Franchise and Pyramid Sales Frauds

If someone approaches you about purchasing a local franchise or distributorship, obtain the names and addresses of franchises near the area where you would operate. Visit them and become familiar with their operations. Determine their successes, ask their opinion of the franchisor, and discuss their problems.

Stay away from any franchise whose major money-making appeal is the sale of other franchises or of lower-level distributorships. Determine whether the franchise has a proven product or service, not just a gimmick or transitory fad. Ask to see certified profit figures of franchises that operate on a scale similar to the one you are contemplating. Check the franchisor's reputation with a better business bureau, and have your attorney review all aspects of the proposed agreement before you sign it.

Security Technique 258: Don't Fall Victim to a Land Fraud

Always read the disclosures in a land developer's property report. There is no law preventing someone from selling you a lot that is swampland, or one that is located far away from a water supply, *if* this is disclosed in the property report. Ask your attorney to review the report and the agreement you need to sign. Do not purchase the property without personally inspecting the site. Do not waive the cooling-off period provided for in contracts. Never base your decisions to purchase on promises. Consider only what is spelled out in the contract you sign.

Preventive Measures Against Embezzlement and Pilferage

A long-known business saying is still true: "The weakest link in your business is your most trusted employee, because he or she is in position to inflict the greatest damage." Keep that advice in mind as you read through the following techniques.

Security Technique 259: Establish Rigid Cash Controls

- Do not allow the department that handles your business accounting to control the receipt and disbursement of funds.

- Support over-the-counter cash sales with serial-numbered sales slips.

- Do not refund merchandise in cash.

- Send refund checks by mail to dissatisfied customers who have a legitimate complaint.

- Do not allow an employee who prepares a deposit slip to make the bank deposit as well.

- Check cash receipts records against duplicate deposit slips.

- Do not allow an employee who balances the day's cash against sales slips or cash register tapes to become involved in sales.

- Depending on the size of your business, consider reconciling your bank statements yourself.

- Examine all canceled checks and endorsements for possible irregularities.
- Prenumber business petty cash slips.
- Keep petty cash separate from other business funds.
- Do not use petty cash funds to cash checks or make loans.

Security Technique 260: Monitor Disbursements

- Pay business bills by check.
- Make sure that payment authorizations are invalidated (e.g., by perforations) after payment so that they cannot be used again to support a fraudulent disbursement.
- Prenumber all blank checks and keep them under lock and key.
- Prepare your checks with a check-writing machine.
- Do not presign checks.
- Do not return signed checks (for mailing) to the person who prepared them.
- Compare canceled checks with invoices or vouchers, noting the names of payees and the dates and amounts.

Security Technique 261: Supervise Accounts Receivable and Sales

- Make sure that correspondence related to accounts receivable is handled by an employee other than the one who prepared the monthly statements.
- Personally approve all adjustments—such as credits for returned goods and pricing errors—or assign the task to a manager who does not have access to cash.
- Assign collection of past-due accounts to someone outside the bookkeeping department.
- Number invoices sequentially.
- Approve bad debt write-offs personally, or assign the task to a manager outside the bookkeeping department.

Security Technique 262:
Manage Purchasing, Receiving, and Accounts Payable

- Separate receiving from purchasing and both from accounts payable.

- Prepare purchase orders on prenumbered forms.

- Send purchase order copies to both receiving and accounts payable.

- Approve invoices for purchased materials only if they are accompanied by the purchase order and the receiving report.

- Do not allow purchasing personnel to be responsible for checking the condition of goods received or for storing and withdrawing inventory.

Checklist for Planning Inventory Controls

The following checklist provides security techniques for making an accurate physical count of business assets. The checklist has been designed in questionnaire form, and it is broad enough to cover most businesses. For a smaller business operation, certain procedures can be eliminated. Others can be emphasized or deemphasized, depending on the inventory items involved or the existence of alternative methods.

Security Technique 263: Develop a Plan for Business Inventories

1. Have the company personnel responsible for conducting the count been identified? Such identification should, at a minimum, name not only the person with overall responsibility for the physical inventory but also those with supervisory responsibility (e.g., by department or location). Auditors and management will then have the names they need to verify questions about the inventory count.

2. Have you made a preliminary tour of inventory locations? Such a tour may suggest problems that could arise during the count or matters that should be covered in the instructions (e.g., new locations where inventory is stored, types of inventory calling for special talents among observers, and areas needing advance preparation).

3. Have you taken steps to:
- Stack, sort, and clean inventory items?
- Segregate defective and obsolete items?
- Identify slow-moving items?
- Identify consigned or other merchandise in stock belonging to others (including material for government contracts)?

4. Have you given consideration to the need for physically counting business inventory that is held by others for processing and storage, or held on consignment or approval?

5. In setting a date for taking the inventory and planning the time sequence of the count, have you given consideration to:
- The timing of the count as it relates to your balance sheet date?
- The best utilization of personnel?
- The needed clearance if overtime is planned (e.g., from labor unions when applicable)?

6. Have you set an alternative date for making the count? An inventory count should be made within a reasonable time before or after the balance sheet date. In evaluating the acceptability of the alternative date, consider the rapidity of inventory turnover and the adequacy of the records supporting the interim changes.

7. In setting a date for the inventory, have you also fixed a specified time for the job to be finished? This is especially critical in a manufacturing operation where departmental operations must be pre-scheduled.

8. Have you anticipated the possibility of overtime for the count crew? Taking a few extra hours to complete a count may be critical if goods are to be released the following day for production or shipment.

9. Have you established procedures for:
- Reporting the progress of the count (especially in large businesses)?
- Changing methods of inventory made during the count?
- Completing the count (in specified areas or in total) before any count records are removed from the inventory items?
- Touring the business (sometimes with auditors) to ensure that all inventoriable items have been counted and that all count records are accounted for?
- Making on-the-spot changes in procedures to accommodate unforeseen conditions and to prevent an uncorrected error from invalidating all or a major part of the inventory?

10. Have you arranged for a member of the auditing staff to be present at preliminary business meetings with employees who are responsible for the count?

Security Technique 264:
Establish Cutoff Procedures

1. Have you established a cutoff date and time for making the count? Cutoff procedures ensure proper correlation of the physical inventory with the accounting records count. To establish a cutoff, you need to institute procedures that stop the flow of materials, products, or merchandise. You must also ensure that transactions are recorded in the proper accounting period in which they happened.

2. Have you made plans for:
   - Closing down production or suspending or controlling operations?
   - Segregating incoming inventory expected to be received during the count?
   - Accounting for inventory that must be shipped out during the count?
   - Controlling the movement of inventory-in-process?

3. Have you alerted the applicable business departments to control prenumbered receiving reports (or supply other means) to identify "prephysical" and "postphysical" shipments of inventory? You will also need to identify and list goods held by vendors for later delivery.

4. If you run a multiplant operation, have you determined whether interplant shipments of goods or materials will be en route on the inventory date? These shipments are not always recorded in the same manner as shipments to outsiders. It is important to decide beforehand which plan you will use to record interplant shipments in your inventory.

Security Technique 265:
Identify the Items to Be
Recorded

Does your count record leave space for the following items?

- Count record number (if not prenumbered)
- Description of the item (including size, number, or other identifying characteristics)
- Quantity and unit of measurement
- Remarks about the condition of the item

- Location (building, floor, department, or section)
- Initials of the count crew (and second crew when applicable)
- Date of the count
- Cross-reference to the inventory summary (for postinventory use)

Security Technique 266: Supervise the Count Record

1. Have the forms for entering the inventory counts been identified? The most widely used forms are preprinted $8\frac{1}{2}'' \times 11''$ sheets on which the counts are recorded directly. Another system relies on separate tags for recording each item. Appropriate supplies are available commercially. In some businesses, counts are recorded on magnetic tape or on calculating machines (with or without printouts).

2. Have you considered using count records of different colors, shapes, or sizes for various categories of inventory?

3. Have you provided for as many sets of count records as will be necessary? A separate set of records may be needed:
- To provide a visible means of determining that all items have been counted
- To provide adequate documentation for the count listing
- To inform the auditor (if applicable and if desired)

4. Have you taken steps to ensure that all items will be counted and no duplicate counts will be made? Three-part tags (or an original and two carbons) serve the multiple purpose of supplying both you and your auditor with a set of count tickets. Because the remaining stubs (or second copy) stay attached to the inventory items, there is a visible means of determining that all items have been counted. When some form of count copy record is not attached to the items, you must find an acceptable alternative procedure.

5. Will your inventory count record be prenumbered?

6. Have you made provisions to control all count records (issued, returned, unused, and spoiled or voided) and to reconcile them after the inventory count?

Security Technique 267: Give Clear Instructions to Your Count Crews

1. Have you used clear terminology in instructing count crews on the nature, condition, and location of the items they will be counting?

You should base the count on *descriptive* characteristics of inventory items, such as their:

- Nature: description, serial, part or model number, manufacturer
- Condition: defective, obsolete, slow-moving
- Location: area of plant or building in which items are kept

2. Did you avoid using similar descriptions for items that have significantly different costs? Not only will similar descriptions become confusing to the count crews, but they are easy to alter deliberately. In one inventory scam, the word "boxes" (meaning empty containers) was altered on the inventory count to "boxed" (meaning containers with the finished product inside).

3. Have your instructions made clear the units of measurement necessary for recording quantities of various inventory items? Quantity, dimension, and capacity can be stated using such measurements as these:

- Units: each, dozen, gross, "lot"
- Weight: avoirdupois, troy, apothecaries, metric
- Dimension: linear, chain, metric
- Capacity: dry, liquid, apothecaries, metric

4. Are your descriptions and units of measurement compatible with those used for costing purposes? Compatibility minimizes the possibility of errors later when pricing information is entered on the count listing.

5. Have you instructed the count crews on the most convenient and reliable methods of counting (consistent with the units of measurement) for various types of inventory items? Sometimes conversion methods can facilitate a count without sacrificing reliability. For example, inventory items can be weighed and then converted to units based on a units-per-pound figure, or rounded off to the nearest pound or foot. Material in piles or in barrels, boxes, or bags stacked in solid formation may call for special equipment or techniques.

Security Technique 268: Record the Inventory Count

1. How will your inventory count be recorded? Recording in ink minimizes later, unauthorized alteration of the count record. Ball-point pens with black ink work best. If a counter or recorder makes an error, a correction should have only one fine line drawn through it. The original price must still be readable, and the corrected figure should be placed next to it and initialed.

2. Will the count be made by "impartial" crews? Count crews should be composed of two employees who are not responsible for custody of the items they count, but are familiar with counting terminology, units of measure, and the manner of measurement.

3. Will a second crew double-check the accuracy of the count record? A self-audit, even on a test basis, improves the accuracy of your inventory count.

4. Do you have methods for reconciling differences if the second crew finds a discrepancy in the count record?

5. Will the initials of the count crew (or some other identification) be placed on the count record?

Security Technique 269: Manage Your Inventory Summary

1. Has the form of the count listing or summary been identified? The count listing summarizes the individual count records. It should provide space for pricing data, dollar extensions, and other identifying comments. The count record may also be designed to serve as the listing.

2. Will your count listing provide for identification (or for separate listings) by categories? Classifications should include:
- Type inventory: raw material, in-process, finished goods
- Condition: defective, obsolete, slow-moving, excess quantities of active items
- Location: building, floor, department or section, off premises, in transit
- Ownership: inventory-in-stock belonging to others
- Source: count record number with cross-reference on the count record to the listing page (particularly important to maintain an audit trail)

Security Technique 270: Use Work-in-Process Inventories

1. Has your inventory count been scheduled at a time when work-in-process will be at a minimum? Inventorying work-in-process can be difficult. However, if the production cycle or other business operation is at its lowest point, the problem is minimized.

2. Is your method of identifying the state of production or operations related to the cost-accounting system? Variations in cost-account-

ing systems prevent other than a general question on this point. In a standard or job cost system, for example, the stage of operations (or percentage of completion) may be identifiable on the basis of manufacturing specification sheets or work orders. In a process cost system, controlled movement implemented throughout enables input and output to become measurable.

3. Will you supply count teams (and your auditor) with specification sheets, work orders, or other documentation to identify stages of production or similar operations? These materials not only must be related to the cost-accounting system but must provide those involved in the count with adequate information to formulate conclusions and record their observations.

Security Technique 271:
Understand the Auditor's Role
in the Inventory Process_____

1. Has guidance been given the auditor's staff regarding the extent of the test checks (and the consequent staff requirements, in terms of both numbers and experience level)? Test checking is influenced by many factors, including the auditor's evaluation of the system of internal control, the thoroughness of your advance preparation and instructions, the need to have representative coverage, and statistical sampling methods (which can affect the extent of the auditor's test checking). In determining representative coverage, you should give consideration to:

- The relative proportion of separate valuable items in the inventory
- Items difficult to count and for which there is a greater chance of error in the count
- Items that, from previous experience, you expect to be significantly book different

2. Have you determined the physical location where the test checks will be conducted? You need to consider deployment of the auditor's staff both within a plant or business and among plants or other locations. A staff member should visit each department or location, including those in which no test checks are made, to see that prescribed procedures are followed.

3. Have you supplied guidance, to both inventory crews and auditors, concerning the use of listing procedures to help trace the audit test checks to the count listing? Catalogs or blueprints may be supplied as reference sources to be used when questions arise. The auditor's working papers must contain enough details of the count records

test-checked to make later tracing into the count listing meaningful. Ordinarily, it is not prudent to initial or mark any of the business's records for this purpose, because that might invite later alteration of records not so marked.

4. Have you established procedures to satisfy the auditor that your business operations have maintained control over movement of goods during the count period? When goods are moved during a count, it is possible to overlook certain items or to double-count them. Such movement may call for the constant presence of an auditor (as an observer) in the receiving or shipping room. You should make a working-paper record of receiving reports, shipping notices, and other documents affecting inventory before, during (if applicable), and after the inventory.

5. Have you given auditing personnel appropriate guidelines so they can count items which they normally would not be familiar with and become acquainted with the manner of measurement (or identification) and the devices used? Unusual problems can arise if an inventory has precious jewelry, fashion merchandise, liquids stored in tanks, or materials stored in piles. Even in ordinary situations, guidance is necessary. For example, you would need to instruct auditing personnel on how to distinguish among cold-pressed steel, hot-rolled steel, and wrought-iron sheets.

6. Have you informed the auditor on whether the devices used to measure (or identify) inventory are accurate? For example, a scale that uniformly reads 10 pounds more than actual weight could materially distort your count. Despite complexities, the auditors must take reasonable precautions to ensure that each test-counted item has the properties attributed to it and that the measurement shown is accurate. In some circumstances, the help of experts may be sought. Auditing personnel should make clear to you, however, that they do not purport to act as appraisers, valuers, or experts on materials.

7. Have you identified the person with whom the auditor can clear problems arising during the inventory count? The auditor's staff needs flexibility to deal with last-minute changes in business inventory instructions or to respond to observations of noncompliance with instructions.

8. Have you alerted the auditing staff to the need to adjust procedures when magnetic tapes or calculating machines are used to make the count? For example, from randomly chosen tapes, your auditor could use a playback machine to transcribe the information to worksheets. Using these worksheets, the auditor would then test-check the

counts. As with written records, some control must be maintained over the tapes in making test counts.

As business owner or manager, you have ultimate responsibility for maintaining the historical record of physical inventory documentation. Outside auditors receiving a complete, duplicate set of count records ordinarily do not keep the set after it has served its purpose.

Appendix H

Criteria for Data-Processing Safeguards

Whenever employee evaluations, trade secrets, cash records, or other sensitive information is kept on computer records, an electronic data processing (EDP) security program is needed. Of course, a sense of proportion is important: Not every piece of confidential information should be guarded like Fort Knox. Base your decision to shield data on two principles:

- What is the maximum amount of damage that would result if the information were revealed?

- How much effort would it take for an outsider (or insider) to gain access to the information and copy, wipe, or falsify it?

The following set of security techniques will help you safeguard your electronic database.

Security Technique 272: Identify Loss Reduction

Determine that a proposed countermeasure adequately addresses a proven threat and lowers the risk.

Security Technique 273: Evaluate Performance Degradation

Determine the price paid for a countermeasure in terms of its direct cost plus the possibility that its use may lower the performance level of the EDP operation.

Security Technique 274: Place Minimum Reliance on Secrecy

Do not base your security plan solely on keeping the functions of a system secret. Often criminals will know as much about the design and implementation of safeguards in a system as the designers of the system themselves. For example, security is not keeping secret how a lock works but rather is safeguarding the keys for the lock. Of course, do not unnecessarily disclose the workings of any system countermeasure.

Security Technique 275: Use the Principle of Least Privilege

This is also the "need to know" principle. To follow it, restrict information about the system's safeguards to the least possible number of employees. These people should know only enough to perform their jobs and to maintain the effective operation of the EDP system.

Security Technique 276: Separate Responsibilities or Use Dual Control

Sensitive functions within the EDP system should be broken down into the smallest effective work assignments. An alternative is dual control over a sensitive function, in which one person performs the function and another oversees the work performed. Accountability for all work must be assured.

Security Technique 277: Ensure Completeness, Consistency, and Reliability

A countermeasure must perform all its specified tasks and functions completely, consistently, and without conflict. It must perform reliably enough to warrant confidence in its continued use. Safeguards must be implemented in a "chain of protection" fashion because security is only as good as its weakest link.

Security Technique 278: Monitor Threats

Carefully craft each countermeasure so that its performance and any threats against its performance (or the assets it is protecting) can be detected and reported. For example, if computer access is protected by requiring users to enter a password, the system should record and report all instances in which incorrect passwords have been entered.

Security Technique 279: Maintain Auditability

A countermeasure must permit an auditor to determine that it is functioning properly and is in compliance with its specifications. For example, your data-processing personnel should be able to demonstrate the ability to recover from a disaster by restating important computer applications from remotely stored copies of backup files and programs.

Security Technique 280: Gain Personnel Acceptance and Tolerance

Employees who are constrained in their work by a countermeasure must be willing to accept the constraint and tolerate its functions and purpose. For example, your programmers must abide by restrictions on access to the computer room.

Security Technique 281: Focus on Sustainability

A countermeasure must function effectively not only when it is first installed but throughout its useful life. For example, plastic covers to protect equipment during emergencies should not be moved or concealed.

Security Technique 282: Compartmentalize

Safeguards must be compartmentalized so that the compromise of one does not lead to the compromise of another. It may be desirable, for example, for your company to restrict physical access to functional areas within the EDP operation as well as on the perimeter.

Security Technique 283:
Isolate Security Safeguards

Security safeguards should be isolated so that personnel constrained by one countermeasure are prevented from compromising or weakening another countermeasure. In essence, employees in positions of trust must function as though they were in a hostile environment. Security becomes vulnerable when an employee is manipulated into cooperating in an unauthorized or illegal act. All employees should be alert to this possibility and be able to resist such an attack.

Security Technique 284:
Account for Legal and Ethical
Constraints

Safeguards must comply with legal and regulatory restrictions. They must not violate the ethical practices of employees or place them in positions of trust beyond their ability to resist temptation. Investigate the background of employees in positions of trust only to the extent warranted by that degree of trust.

Appendix I

Physical Security Plan for Businesses

It is essential to your business's prosperity that you develop a detailed physical security plan. The plan should include access and asset control, protective barriers and lighting systems, and other important devices. Update your security plan regularly to meet changing business environments and needs.

Matrix for Creating a Physical Security Plan

The following techniques supply you with the necessary matrix to create a physical security plan tailored to your business needs.

Security Technique 285: State Your Purpose

Define the purpose of your security plan.

Security Technique 286: Determine Area Security

Define the areas, buildings, and other structures considered critical and establish priorities for their protection.

Security Technique 287:
Establish Control Measures_____

Establish restrictions on access to and movement within critical areas. These restrictions can be categorized as personnel, material, and vehicle control.

Personnel (Employee, Visitor, Vendor) Control

1. Access controls pertinent to each area or structure
    a. Authority for access
    b. Access criteria for
        (1) Business employees
        (2) Visitors
        (3) Vendors
        (4) Maintenance people
        (5) Contractors
        (6) Others
2. Identification
    a. Describe the system used in each area. If you use a badge identi-fication system, supply a description covering all requirements of identification and control of people working or conducting business within your premises.
    b. Application of the system
        (1) Business personnel
        (2) Visitors to restricted areas
        (3) Visitors to administrative areas
        (4) Vendors
        (5) Contractors
        (6) Maintenance people
        (7) Others

Material Control

1. Incoming
    a. Requirements for admission of material and supplies
    b. Search and inspection of material for possible hazards
    c. Special controls for delivery of supplies or other shipments into restricted areas
2. Outgoing
    a. Documentation needed
    b. Controls
    c. Sensitive shipments (when applicable)

Vehicle Control

1. Policy on business or company vehicles
2. Policy on employee vehicles

3. Policy on vendor and visitor vehicles
4. Controls for entrance into restricted and administrative areas
   a. Employee vehicles
   b. Business or company vehicles
   c. Emergency vehicles
   d. Vendor, visitor, contractor, and other vehicles
5. Policy and procedure for vehicle registration (in industrial or manufacturing operations)

Security Technique 288: Use Security Aids

Show the manner in which the following aids to security will be implemented within business areas.

Protective Barriers

1. Definition (fences, buildings, railings, etc.)
2. Clear zones (areas free from weeds, shrubs, etc., around fences and buildings)
   a. Criteria or needs
   b. Maintenance
3. Signs (instructional signs for personnel and vehicle control)
   a. Types
   b. Posting
4. Gates (for personnel and vehicle control)
   a. Hours of operation
   b. Security requirements
   c. Lock security

Protective Lighting System

1. Use and control (internal and external)
2. Inspection
3. Action you will take if power failure happens
4. Action you will take if alternative source of power fails
5. Emergency lighting systems
   a. Stationary
   b. Portable

Intrusion Detection Systems

1. Security classification
2. Inspection
3. Use and monitoring

4. Action you will take (or event that needs to happen) if your intrusion alarm activates

5. Maintenance

6. Alarm logs or registers

7. Sensitivity settings

8. Fail-safe and tamperproof provisions

9. Monitor panel location

Communications

1. Location

2. Use

3. Testing

Security Technique 289: Instruct Security Personnel

Include general instructions that would apply to all security officers (fixed and mobile). Detailed instructions on policy and procedures should be attached as annexes.

1. Number of security officers
2. Length of shifts
3. Essential duties, posts, and routes
4. Weapons and equipment
5. Training
6. Use of sentry dog teams (for added security in high-crime or plant areas)
7. Reserve security forces (when applicable)
   a. Composition
   b. Purpose or mission
   c. Weapons and equipment
   d. Location
   e. Utilization concept

Security Technique 290: Set Contingency Plans

Describe actions you need to take in response to various emergency situations. Attach as annexes detailed plans for responding to bomb threats, disasters, fire, robberies, and burglaries.

1. Employee actions
2. Police response
3. Security officer actions

Security Technique 291: Coordinate All Efforts

Describe matters that call for coordination with outside agencies (fire departments, law enforcement personnel, and others).

1. Integration with certain plans and needs of agencies
2. Liaison and coordination
   a. Local authorities
   b. Local support elements

Security Technique 292: Provide Appropriate Supplementary Material

Add any other information you believe to be important to your security plan. Examples include detailed contingency plans to respond to business emergencies.

Negotiating with a Private Security Company

Security personnel are supplements to—not replacements for—a sound physical security plan. Establish a solid program of deterrents and preventive measures, along with employee training. If you still need security assistance—for example, if you are in a high-crime area—enlist the services of a reputable agency. You need to negotiate the following points.

Security Technique 293

Determine what the security company will pay its security officer at your business compared with what you must pay the security company for the service. Obviously, the security company must have a profit; however, frequently the differences will prove excessive.

Security Technique 294

Determine if the contractual fee that you pay a security company also includes money set aside to supply the security officer with pay increases and other benefits.

Security Technique 295

Ask for a complete accounting of the security company's hourly rate above that paid to the security officer.

Security Technique 296

Verify the qualifications and training requirements of security officers.

Security Technique 297

Obtain proof of the insurance coverage carried by the security company and notice of any insurance cancellation.

Security Technique 298

Find out how many hours each week the officer will work and if the security company pays overtime rates.

Security Technique 299

Make certain you have the right to cancel the security contract on 30 days' written notice.

Security Technique 300

Retain the right to have any officers removed from your business at your discretion.

Security Technique 301

Ensure that the responsibilities assigned to a security officer at your business meet with your approval and create genuine benefits to your operation.

Index

Advertising:
    deceptive, 201–203, 205
    and sensitive business information,
        226
Air conditioners:
    alarms on openings, 87
    space beneath, 7
Air ducts, 7, 46
    alarms on, 87
Air transport, 248–249
Alarm detection systems, 92–94
Alarm glass, 74
Alarm screens, 75, 87
Alarms, 90
    on doors, 2, 5, 15, 46–49, 86
    power source, 58
    on windows, 2, 4, 74–75, 86–87
    (*See also* Intrusion detection systems)
Alcoholics:
    and robbery, 98
    and shoplifting, 178
American Bankers Association routing
        symbol, 141–145
Antitrust violations, 205–214
    franchise tying agreements, 209–210
    interlocking directorates, 210–211
    market share, 208–209
    mergers and acquisitions, 210
    monopolies, 207–208
    price discrimination, 211–212
    relevant market, 208
    remedies and enforcement, 212–214
    restraint of trade, 206
    rule of reason and per se doctrine,
        206–207
    tying agreements, 209–210

Armed robbery, 10, 95–111
    anatomy of, 98–103
    assessment of risk of, 103–106
    business of, 97, 99–100, 103–105, 107
    defined, 96
    and intrusion detection systems, 68
    offenders, described, 96–98
    prevention of, 107–109
    response to, 100, 106–107, 109–110
    and security officers, 285
    theft versus, 96
Automatic telephone dialer alarm
        systems, 72, 90–91

Background checks, 261–262
Bait-and-switch supply schemes, 199–201
Banks:
    armed robbery, 104
    bogus deposit slips, 231
    disclosure of business information, 222
    vaults, 2
Basement windows, 5
Better business bureau:
    and coupon fraud, 166–168
    and preemployment screening
        operations, 255
Burglary, 2, 10
    and intrusion detection systems, 68
Business format franchises, 209–210
Business information, 12–14, 216–236
    classification of, 219–220
    and computers, 225, 227–235
    embezzlement of, 219, 229–235
    and fax machines, 226–228
    internal security measures, 223–227

Business information (*Cont.*):
  locks and keys for sensitive, 37–42,
    223–224
  and mobile offices, 227–228
  pilferage of, 118
  and public places, 221–223
  security of, 217–220
  and telephones, 217–219, 227–228
  theft of, 28, 225, 229–235
Business-to-business fraud, 194–215
  antitrust violations, 205–214
  bait-and-switch supply schemes,
    199–201
  check-writing, 229–230
  coupon scams, 166–168, 170–171
  duplicate-billing schemes, 197–199
  gray market, 194–197
  insurance, 230–231
  unfair competition, 201–205

Capacitance proximity detectors, 75
Cargo:
  diversion of, 237–252
  protective management of intransit
    cargo, 238–243, 245–246, 249–250
  ripple effect of diversion, 244–246
  techniques to prevent diversion, 250–251
  transportation mode, 246–250
  and verification of shipments, 239
Cash (*see* Checkout counters; Counterfeit
  currency; Discount coupons)
Cashier's checks, 150
Ceilings, 8
  alarms on, 86
  drop, 8
  hardening, 55
Cellular telephones, 227–228
Chamber of commerce, and coupon fraud,
  166–168
Checkout counters, 11–12
  and coupon fraud, 169–170
  and employee pilferage, 116, 117,
    133–134
  and shoplifting, 183, 190–191
Checks, 136–150
  business, 140, 142–148
  cashier's, 148, 150
  check-writing fraud, 229–230
  courtesy check-cashing cards, 139, 142
  and credit card information, 133, 137

Checks (*Cont.*):
  forged, 140–142
  profile of bad check artist, 136–140
  traveler's, 148–150
  verification procedures, 133, 138–140
Children and shoplifting, 176, 180, 183
Civil Aeronautics Board (CAB), 205
Class actions, 213
Clayton Act, 209, 210, 212, 213
Closed-circuit television (CCTV), 108, 175,
  183, 186
Combination locks, 21–23
  manipulation-resistant, 23
  pros and cons of using, 25–28
  relocking devices for, 23
  security measures for, 34–35, 40
Competition, unfair methods of, 201–205
Computers and business information, 225,
  227–235
Confidential information (*see* Business
  information)
Consent decrees, 212–213
Construction openings, alarms on, 87
Consultants, 283–284
  and confidentiality, 293
  and consulting agreements, 296–297
  described, 283–284
  determining need for, 292–293
  fees of, 292–293
  proposals from, 294–296
Consumer Product Safety Agency, 205
Containers:
  control of incoming, 16
  (*See also* Cargo, diversion of)
Continuous lighting (stationary
  luminary), 59–60
Controlled lighting, 60
Convex mirrors, 175
Core-lock systems, 15, 23–24
Counterfeit currency, 154–163, 172–173
  circulation of, 162–163
  creation of, 159–162
  mechanics of, 154–156
  types of, 156–159
Coupons (*see* Discount coupons)
Crash bar doors, 15, 46–49
Credit bureau reports, 256–259
Credit cards, 150–152
  common scams, 151–152
  and employee pilferage, 132–134
  verification procedures, 150–152

Customers:
  employee pilferage against, 132–134
  shoplifting by (*see* Shoplifting)
Cylinder locks, 18
  described, 19–20
  pros and cons of using, 29–31
  security measures for, 36
Cypher locks, 24

Data transmission alarm systems, 91–92
Dead-bolt latches, 29–30, 33, 34, 37, 39
Department of Agriculture, 204
Deposit slips, fake, 231
Discount coupons, 163–171
  coupon kings and queens, 168–169
  and employee pilferage, 169–170
  fake, 165, 166
  scams involving, 166–168, 170–171
Disk tumbler locks, 20–21
Display cases for hardening walls,
    52–55
Display counters, 11, 12
Doors, 4–6, 44–49
  alarms, 2, 5, 15, 46–49, 86
  crash bars, 15, 46–49
  electronic article surveillance (EAS)
      systems, 191
  frames, 5, 33, 44–45
  roof, 7
  for sensitive storage areas, 37–42
  transoms, 5–6, 46
  windows on, 4
  (*See also* Locks)
Doppler sensors, 76
Drop ceilings, 8
Drug users:
  and robbery, 98
  as shoplifters, 178
Ducts, 7, 46
  alarms on, 87
Dun & Bradstreet, 222
Duplicate-billing schemes, 197–199
Duress alarms, 75, 88, 106–107

Electronic article surveillance (EAS)
    systems, 191
Elevators, 46
Embezzlement of business information,
    219, 229–235

Emergency lighting, 58, 60–61
Employees:
  and cargo diversion, 238–239, 243
  embezzlement by, 219, 229–235
  hiring (*see* Hiring process)
  nondisclosure agreements, 226
  pilferage against business customers,
      132–134
  temporary, 243
  training of (*see* Training employees)
  (*See also* Pilferage)
Employment applications, 259–262
Environmental Protection Agency (EPA),
    205

Facsimile machines, 226–228
False alarms and intruder detection
    system, 71–72
Federal Communications Commission
    (FCC), 204
Federal Deposit Insurance Corporation
    (FDIC), 205
Federal Fair Credit Reporting Act,
    256
Federal Home Bank Board, 205
Federal Reserve Board, 205
Federal Reserve code:
  on checks, 141–143, 145–147
  on currency, 157–158
Federal Reserve notes, 156–158
Federal Trade Commission (FTC),
    202–203, 212
Floors, alarms on, 86
Fluorescent lamps, 57
Food and Drug Administration (FTC),
    203–204
Forged checks, 140–142
Frames, door, 5, 33, 44–45
Franchise tying agreements, 209–210
Fraud (*see* Business-to-business fraud)

Gaseous discharge lamps, 57
Gate lock, 25, 26
Glare projection lighting, 59
Goldfarb v. *Virginia State Bar*, 207
Gray market, 194–197
  defined, 195
  mechanics of, 195–196
  protection from, 197

Hatchways, 46
Hiring process, 253–266
    background checks in, 261–262
    for consultants, 283–284
    employment applications in, 259–262
    interviews in, 264–266
    preemployment screening in, 254–259
    for security officers (*see* Security
        officers)
    testing in, 262–264

Incandescent lamps, 57
Indirect theft, 115–116
Information (*see* Business information)
Information brokers, 221, 225, 228,
    254–255
Infrared motion sensors, 76
Insurance fraud, 230–231
Intangible assets (*see* Business
    information)
Interstate Commerce Commission (ICC),
    213
Interviews in hiring process, 264–266
Intruder time requirement, 70
Intrusion detection systems (IDS), 65–94
    advantages of, 66–67
    described, 65–66
    factors affecting selection of, 67–70
    and false alarms, 71–72
    parts of, 70–71
    problems with, 71–74
    types of, 74–94
    (*See also* Alarms)
Inventory:
    accountability and control, 126–132,
        240–243
    and prevention of cargo diversion,
        240–245
    and shoplifting (*see* Shoplifting)
    in warehouse operations, 12, 13

Job descriptions, 260–261

Keys:
    control system for, 14–15, 35–42
    high-security, 37, 39–41
    for sensitive information/property,
        37–42

Keys (*Cont.*):
    skeleton, 21
    (*See also* Locks)
Kleptomaniacs, 178

Laser sensors, 76
Layout and shoplifting opportunities,
    182–186
Lighting:
    emergency, 58, 60–61
    exterior, 8–9, 56–64
        above doors, 45
        maintenance and inspection,
            61–63
        selection of, 62–63
        types of, 57, 59–60
    interior, 56
        and shoplifting opportunities,
            182–186
        for training program, 281
    power sources for, 58
    wiring for, 58
Lock breaks, 29, 31–33
    loiding, 29
    rapping, 28
    sneaker, 28
Locks, 18–43
    on cargo shipments, 245–246, 250–251
    combination, 21–23, 25–28, 34–35, 40
    core-lock system, 15, 23–24
    cylinder, 18–20, 29–31, 36
    cypher, 24
    dead-bolt latches, 29–30, 33, 34, 37, 39
    disk tumbler, 20–21
    door, 5, 45
    high-security, 37–42
    key control system, 14–15, 35–42
    lock and key systems, 14–15, 20, 33–42,
        223–224
    maintenance and inspection procedures
        for, 41–42
    picking, 29, 31–33
    pin tumbler, 18–20, 29–31, 36
    quality of, 5, 15
    for sensitive information/property,
        37–42
    wafer, 20–21
    warded, 21, 22, 28–29
    window, 45
Loiding, 29

Management controls, 13–16
    in hiring (*see* Hiring process)
    over incoming packages and containers, 16
    and intangible assets, 12–14
    lock and key system, 14–15, 20, 33–42, 223–224
    in training (*see* Training employees)
    over vehicle access, 16
    in warehouse operations, 12
    (*See also* Business information)
Manufacturing operations, 10, 12–13
    cargo diversion (*see* Cargo, diversion of)
    pilferage from, 117, 119–120
    and security officers, 285
    vehicle access in, 16
Market share, 208–209
Master brand combination locks, 25
Meetings, confidentiality of, 221–223
Mercury vapor lamps, 57
Mergers and acquisitions, 210
Metal halide lamps, 57
Metallic foil sensors, 75, 77
Microwave motion sensors, 76, 85
Mirrors, 175, 186
Monopolies, 207–208
Motor transport, 246–247

Nondisclosure agreements, 226

Office operations, 10
    pilferage from, 120–121
Office supply scam, 199–201
Overbilling schemes, 197–199

Packages:
    control of incoming, 16
    (*See also* Cargo, diversion of)
Paper shredders, 133, 225
Parking arrangements, 16
Per se doctrine, 206–207
Photoelectric controls, 75, 78–82
Photoelectric sensors, 75
Pilferage, 10, 112–135
    and basement windows, 5
    business customers, 132–134
    of cargo (*see* Cargo, diversion of)
    categories of pilferers, 122–123

Pilferage (*Cont.*):
    and coupon fraud, 169–170
    indirect, 115–116
    and intrusion detection systems, 68
    inventory accountability and control, 126–132
    magnitude of, 113–121
    in manufacturing operations, 119–120
    in office operations, 120–121
    reasons for, 121–126
    in retail operations, 114–118
    in warehouse operations, 118–119
Pin tumbler locks, 18, 29–31
    described, 19–20
    pros and cons of using, 29–31
    security measures for, 36
Point sensors, 88–92
Police:
    and risk of armed robbery, 103–104
    and shoplifting, 188–189
    working with, to prevent crime, 107–108
Polygraph tests, 263
Power sources:
    backup, 58, 60–61, 73–74
    for intruder detection systems, 73–74
    for lighting, 58
Preemployment screening:
    critical steps in, 256–259
    by outside firms, 254–255
Pressure mat sensors, 75
Price fixing, 207
Professional armed robbers, 97, 99–100, 103–105, 107
Professional shoplifters, 176–177, 186–188
Proprietary information (*see* Business information)
Pull-trip switches, 75

Rail transport, 247
Rapping, 28
Relevant market, 208
Response time:
    and armed robbery, 100
    to intrusion detection system, 69–70, 72
Rest rooms, drop ceilings in, 8
Restraint of trade, 206
Retail operations, 10–12
    cargo diversion (*see* Cargo, diversion of)
    checks in (*see* Checks)

Retail operations (*Cont.*):
    and counterfeit currency (*see*
        Counterfeit currency)
    credit cards in (*see* Credit cards)
    discount coupons in (*see* Discount
        coupons)
    layout of, 10–12, 183–186
    pilferage from, 114–118, 133–134
    shoplifting from (*see* Shoplifting)
    vehicle access in, 16
    (*See also* Armed robbery)
Return merchandise scam, 186–188
Robbery (*see* Armed robbery)
Robinson-Patman Act, 211–212
Roofs, 6–7
    hardening, 55
Rule of reason, 206–207

Securities and Exchange Commission
    (SEC), 202
Security officers:
    armed, 286–287, 290–291
    authority of, 287
    and citizen's arrest, 287–288
    contracting for private, 288–291
    described, 283–284
    determining need for, 284–285
    effective use of, 286
    hiring your own, 291
Senior citizen shoplifters, 177–178
Sensor devices, 70, 74–77
    compatibility of, 77
    redundancy of, 78
    types of, 74–76
Serial numbers, combination lock, 25
Sherman Antitrust Act, 205–207, 210–213
Shimming, 21, 28, 29
Shoplifting, 10, 174–193
    accomplices in, 176–177, 180–181
    apprehending shoplifters, 188–190
    assessment of opportunities for,
        181–186
    and children, 176, 180, 183
    and employee training, 182, 186,
        189–190
    methods of, 179–181, 186–188
    perpetrators, 175–178
    prevalence of, 174
    return merchandise scam, 186–188
    and security officers, 285

Shoplifting (*Cont.*):
    store layout and atmosphere, 182–186
    and underpayment for merchandise,
        180, 190–191
Shredders, 133, 225
Signs on windows, 5
Skeleton keys, 21
Skylights, 7, 45–46
Sneaker device, 28
Sodium vapor lamps, 57
Sound-monitoring sensors, 76, 82
Standard Oil Company v. *United States*, 206
Standby lighting, 60
Stock/storage rooms:
    high-security keys and locks for, 37–42
    and prevention of cargo diversion,
        240–243
    in retail operations, 11, 12
Stress detectors, 75
Structural security, 2–9, 44–64
    ceiling, 8, 55, 86
    doors, 4–6, 44–49
    and intrusion detection systems, 68
    lighting, 8–9, 45, 56–64, 182–186
    other openings, 7, 45–46
    roof, 6–7, 55
    walls, 2–3, 50–55, 86
    windows, 2, 4–5, 45, 46, 48, 86–87
Switch sensors, 75–76, 86

Teamwork:
    and hiring process, 265–266
    and training process, 276–277
Telemarketing methods, 200
Telephone dialer alarm systems, 72, 90–91
Telephones:
    cellular, 227–228
    and security business information,
        217–219, 227–228
Temporary employees, 243
Testing:
    in hiring process, 262–263
    in training process, 276, 280
Theft, 10
    robbery versus, 96
    underpaying for merchandise, 180,
        190–191
Training employees, 267–282
    and business information, 217–219,
        221–223, 226

Training employees (*Cont.*):
    importance of, 268–269
    objectives of, 270–272
    positive environment for, 280–281
    program development, 269–278
    program evaluation, 270–271, 278–280
    and shoplifting, 182, 186, 189–190
    and teamwork, 276–277
    testing in, 276, 280
Transoms, 5–6, 46
Travel plans, confidentiality of, 221–223
Traveler's checks, 148–150
Treasury Department, 205
Truth-in-Lending Law, 205
Tying agreements, 209–210

Ultrasonic motion sensors, 76, 83–84, 87,
    89
Ultraviolet (UV) light:
    and counterfeit bills, 163
    and counterfeit coupons, 165, 166
    and credit card fraud, 152
Uniform Crime Report, 103–105
United States v. *E. I. Du Pont de Nemours &*
    Company, 208
U.S. Postal Service, 204

Vagrants, 178
Vaults:
    combination locks for, 23, 40
    walls of, 2
Vehicles:
    access to business, 16
    in armed robberies, 100, 101, 103
    control over access by, 16

Vehicles (*Cont.*):
    high-security keys and locks for,
      38–42
    mobile offices in, 227–228
Vendors:
    and cargo diversion (*see* Cargo,
      diversion of)
    duplicate-billing schemes, 197–199
    and employee pilferage, 115
    and gray market, 194–197
    office supply scam, 199–201
Vents, 7, 46
Vibration sensors, 76, 82–83, 87

Wafer locks, 20–21
Wafer switch sensors, 76
Walls:
    alarms on, 86
    exterior, 2–3, 50–55
Warded locks:
    described, 21, 22
    pros and cons of, 28–29
Warehouse operations, 10, 12, 13
    cargo diversion (*see* Cargo, diversion of)
    pilferage from, 118–119
    and security officers, 285
    vehicle access in, 16
Water transport, 248
Windows, 4–5
    alarms on, 2, 4, 86–87
    basement, 5
    hardening of, 45, 46, 48
    locks on, 45
    signs in, 5
    (*See also* Locks)
Wiring for lighting systems, 58

About the Author

Russell Bintliff worked as a special agent investigating
white-collar crime, and had more than three decades
of experience in law enforcement and intelligence operations
for the federal government. He was both a small-business
owner and a security consultant to Fortune 500 companies.
Mr. Bintliff authored several books on security, including
The Complete Manual for Corporate and Industrial Security.